Risk and Predictive Analytics in Business with R

Özgür M. Araz and David L. Olson

CRC Press
Taylor & Francis Group
Boca Raton London New York

CRC Press is an imprint of the
Taylor & Francis Group, an **informa** business
A CHAPMAN & HALL BOOK

First edition published 2026
by CRC Press
2385 NW Executive Center Drive, Suite 320, Boca Raton FL 33431

and by CRC Press
4 Park Square, Milton Park, Abingdon, Oxon, OX14 4RN

CRC Press is an imprint of Taylor & Francis Group, LLC

ISBN: 9781032912691 (hbk)
ISBN: 9781032912875 (pbk)
ISBN: 9781003562399 (ebk)

DOI: 10.1201/9781003562399

Typeset in Palatino
by Newgen Publishing UK

Risk and Predictive Analytics in Business with R

Supply chain operations face many risks, including political, environmental, and economic. The past five years have seen major challenges, from pandemic, impacts of global warming, wars, and tariff impositions. In this rapidly changing world, risks appear in every aspect of operations. This book presents data mining and analytics tools with R programming as well as a brief presentation of Monte Carlo simulation that can be used to anticipate and manage these risks. RStudio software and R programming language are widely used in data mining. For Monte Carlo simulation applications we cover Crystal Ball software, one of a number of commercially available Monte Carlo simulation tools.

Chapter 1 of this book deals with classification of risks. It includes a typical supply chain example published in academic literature. Chapter 2 gives a brief introduction to R programming. It is not intended to be comprehensive, but sufficient for a user to get started using this free open source and highly popular analytics tool. Chapter 3 discusses risks commonly found in finance, to include basic data mining tools applied to analysis of credit card fraud data. Like the other datasets used in the book, this data comes from the Kaggle.com site, a free site loaded with realistic datasets.

The remainder of the book covers risk analytics tools. Chapter 4 presents R association rule modeling using a supply chain related dataset. Chapter 5 presents Monte Carlo simulation of some supply chain risk situations. Chapter 6 gives both time series and multiple regression prediction models as well as autoregressive integrated moving average (ARIMA; Box-Jenkins) models in SAS and R. Chapter 7 covers classification models demonstrated with credit risk data. Chapter 8 deals with fraud detection and the common problem of modeling imbalanced datasets. Chapter 9 introduces Naïve Bayes modeling with categorical data using an employee attrition dataset.

Features:

- Overview of predictive analytics presented in an understandable manner
- Presentation of useful business applications of predictive data mining
- Coverage of risk management in finance, insurance, and supply chain contexts
- Presentation of predictive models
- Demonstration of using these predictive models in R
- Screenshots enabling readers to develop their own models

The purpose of the book is to present tools useful to analyze risks, especially those faced in supply chain management and finance.

CHAPMAN & HALL/CRC Series on Statistics in Business and Economics

Recently Published Titles

Empirical Research in Accounting
Tools and Methods
Ian D. Gow and Tongqing Ding

Risk and Predictive Analytics with R
Özgür M. Araz and David L. Olson

For more information about this series, please visit:

Contents

About the Authors

Özgür M. Araz is the Ronald and Carol Cope Professor and Professor of Supply Chain Management and Analytics at the University of Nebraska-Lincoln. His research interests are systems simulation, business analytics, healthcare operations, and public health informatics.

David L. Olson is the James and H.K. Stuart Chancellor's Distinguished Chair in the Department of Supply Chain Management and Analytics at the University of Nebraska-Lincoln. His research interests are data mining, knowledge management, multiple criteria decision-making, and simulation modeling.

1

Measuring and Managing Risk

1.1 Natural Disasters

Earthquakes, floods, fires, and hurricanes are natural disasters. Recent events include the 2020 COVID-19 pandemic; on 6 February 2023, a major earthquake hit Southern Turkey (killing over 50,000) and Syria (killing over 8,000); on 13 March 2023 Mount Merapi volcano in central Java erupted, covering the vicinity in hot volcanic ash; Typhoon Doksuri flooded over 16 cities in northeastern China on 29 July 2023. Nature provides many benefits, but also can be occasionally vicious. To cope with natural disasters requires preparation. The natural science perspective can provide some foreknowledge of disaster, focusing on environmental, economic, and technical systems.

1.1.1 Measuring Natural Risk

There are a number of agencies that have developed metrics for the natural risk factor of climate change (Feldmeyer et al., 2021). These include:

- The INFORM Index developed by the European Union's Joint Research Center ranking countries at risk to climate change and natural hazards. Key dimensions measured are hazard, exposure, vulnerability, and coping capacity.
- The WorldRiskIndex is a model combining physical and spatial exposure to natural hazards with societal vulnerability. The index consists of 28 indicators with the three components of susceptibility, coping capacity, and adaptive capacity. Visualization and communication are key emphases.
- The Global Climate Risk Index is an annual measure of countries and regions impacted by weather-related loss events. It aims to show vulnerability, focusing on the effects of past events.

DOI: 10.1201/9781003562399-1

- The Notre Dame Global Adaptation Index is annually published by the University of Notre Dame, ranking country vulnerabilities with respect to climate change and readiness to adapt. It seeks to provide decision-makers with data to enable them to better prioritize investments and increase resilience. Measures include social, economic, and governance dimensions.

Thus, there are a number of efforts to provide data to decision-makers with respect to natural risks, especially in the domain of climate change.

1.1.2 Managing Natural Risk

Once risks are identified, risk management needs to develop programs to reduce likelihood of their occurrence and minimize their impact. On a macro level, weather and climate disasters threaten life and property throughout the world. Risk management is focused on prediction of rare events and investment in systems to cope with extreme events. Preparation is emphasized, along with development of system resilience (Forrest and Milliken, 2018). This report led to the formation of the Resilient America Roundtable to aid decision-makers in building more resilient systems. Recommendations were to:

1. Communicate, understand, and manage risk
2. Share data and information about hazards, best practices, communications, and policies
3. Measure resilience
4. Build coalitions and partnerships across public, private, NGO, and academic stakeholders

The US National Academy of Sciences provided recommendations suggesting strategic steps that could improve disaster resilience, which was defined as the ability to prepare and plan for, absorb, recover from, or more successfully adapt to actual or potential adverse events (National Research Council, 2012). As an example, the Roundtable applied pilot studies in measuring resilience against flooding in Charleston, SC, and in Cedar Rapids, IA. A framework was obtained from Zurich Alliance insurance firms to identify flood resilience strengths, challenges, and priorities, followed by identification of actions useful in building resilience to floods and to measure progress. Community assets in human (education, health), social (networks), physical (infrastructure), natural (ecosystem services), and financial (diversity, institutions) resources were identified. This study led to the identification of several common challenges. Both Charleston and Cedar Rapids measured robust approaches to watershed management, community engagement, and improving preparedness for flood risks. Resilience priorities were established, and actions identified to address those priorities.

1.2 Malicious Acts

Malicious acts are intentional on the part of fellow humans who are either excessively competitive or who suffer from character flaws. Examples include terrorist activities in Somalia, Syria, and Ukraine/Russia. Criminal activities such as product tampering or kidnapping and murder are clearly not condoned. Malicious acts include business activities such as theft, fraud, vandalism, and corporate espionage. Malicious activity has even arisen within the area of information technology, in the form of identity theft, ransomware, and computer viruses.

Malicious events are created by human action. Wars fall into that category, but while wars continue to be present, hopefully they will not reach the magnitude of the major wars of the 20th Century. Conversely, while terrorism has always been present, it has been especially notable in the 21st Century. Another form of malicious event that impacts business is cyber terrorism, or on a less drastic level, actions such as identify theft and hacking. Risk management against war and terrorism usually involves government action, often involving massive expenses incurred to counter real or imagined threats. Businesses get more involved in countering cyber-attacks, again incurring massive expenses in efforts to thwart hacking activity.

The third type of risk is systemic risk, which we will discuss in greater detail.

1.3 Systemic Risk

Systemic risk involves rippling effects among interconnected agents. In the financial domain, a great deal of effort has been given to measure portfolio performance in terms of expected shortfall or value at risk.

To manage systemic risks, they need to be accurately measured and modeled to include both static structure of networks as well as dynamic processes operating on this structure. The process of contagion applies to diseases as well as computer viruses. In power grids, load redistribution is a crucial dynamic generating systemic risk. In financial networks, one bank defaulting may precipitate default of other banks. Supply chains are also subject to contagion. Agent-based modeling and game theory have been used in attempts to model these kinds of systems (Hochrainer-Stigler et al., 2020).

1.3.1 Defining Systemic Risk

Complex adaptive systems are viewed as having a life of their own, referred to as autopoiesis. Autopoietic systems can create and maintain themselves. Supply chains often consist of independent agents that might participate in all parts of a supply chain temporarily, with new sources entering through market competition.

Complex adaptive systems exhibit emergent behavior. Phenomena emerge from a collection of interacting objects in a complex, non-linear fashion evading precise modeling. System development follows an arrow of time, in that it is one way, irreversible. System elements interact through feedback. Looking at systems can lead to understanding at the macro-level which can lead to better prediction and control.

Complex adaptive systems include the following components:

- Elements – agents, with different degrees of autonomy, interaction, and learning
- Behaviors – co-evolution and self-organization when faced with challenges
- Effects – adaption that is often non-linear and irreversible

Table 1.1 gives some elements of complex adaptive systems (CAS) taken from two studies (Olson, 2015 on a supply chain; Olson, 2017 on the 2008 real estate market).

The field of systemic risk that has received the greatest attention is the financial sector. That field provides the most developed set of risk measures and management tools.

1.3.2 Managing Systemic Risk

The most common means to manage systemic risk is insurance, which is highly appropriate for common events such as automobile accidents. Insurance is a form of diversification, spreading risk liability over a broader set of entities. This approach is less effective for extreme-event risks, which can have ripple effects if a major natural catastrophe overwhelms an insurance company's resources. An approach to mitigate these extreme-event risks is structural diversification, which may enable risk sharing and facilitate post-failure recovery. An example might be the Federal Emergency Management Agency in the US, to include government resources applied in disaster areas. A key focus of reshaping a network's topology to minimize systemic risk would be

TABLE 1.1

Complex Adaptive System Elements

Systems Concept	CAS Element	Supply Chain	Housing System
Autopoiesis			Bank deregulation CDO evolution
Internal mechanism	Agents	Countries Companies	Home borrowers Mortgage lenders Investment banks Rating agencies Investors Insurers
"	Self-organization & emergence	Formation of cartel	Bank deregulation E-lending
"	Connectivity	Mine control Electricity supply	E-business Government bailout
"	Dimensionality	Feedback Australian ore	Moral hazard
Environment	Dynamism	Mine exhaustion Taxation policy	Housing prices
Co-evolution	Non-linear change	Unexpected consequences	Loan regulation Rating agencies

to identify nodes that are too big or interconnected to fail, as in investment banks in 2008. However, diversification strategies can have unintended and undesirable consequences. The human-agency aspect of the system needs to be included. Features characterizing differences in types of systemic risk include indetermination, indecision, and responsibility.

Variability due to free will is a fundamental source of indetermination. If human interactions are involved, systemic risk is more unpredictable, harder to define, and involves more outlying events. Collective indecision generated by mutual uncertainty leads to suspension or alternation of rules and procedures, and thus systemic risks cannot predict outcomes. Human choice defies mathematical prediction. Analogies from natural sciences such as the concept of contagion downplay human agency and responsibility falsely leading to conclusions that systemic risk is only technical.

1.3.3 Micro Example

On a micro level, Abramov and Al-Zaidi (2023) developed criteria for construction firm risk management. These criteria were grouped into economic, organizational/technical, political/military, and occupational safety categories. While specifically developed for construction project management in Iraq, for the most part they apply to any construction project.

In the economic sector, criteria for reducing or limiting the impact of risk involved:

1. Initial assessment of partner bankruptcy
2. Monitory government policy, inflation, and tax regimes
3. Make necessary bank arrangements for project financing
4. Manage exchange rate fluctuations with respect to materials, equipment, and labor
5. Avoid corruption and bribery
6. Develop a reserve of funding to deal with possible price increases
7. Attract foreign investment to ensure competition

In the organizational/technical field:

1. Develop communication channels and information transfer across stakeholders
2. Develop a change-order process to ensure compliance with technical standards
3. Identify project activity deadlines
4. Provide for skilled labor
5. Use modern technology and procedures
6. Timely maintenance of machines and equipment
7. Plan for deviation from project schedule from unforeseen events
8. Review of equipment and materials suppliers

In regard to political/military decisions:

1. Monitor emergency situations in the construction zone
2. Develop alternative routes of delivery
3. Post warning signs
4. Installation of video camera and security checkpoints

Occupational safety factors included:

1. Providing fire-fighting equipment and emergency ladders
2. Conducting safety briefings and training
3. Providing personal protective equipment
4. Provide for high-temperature working protection
5. Insuring workers and equipment

While some of these factors were specific to Iraq, we have included those that appear generally applicable to construction project risk management.

1.4 Conclusions

This chapter has presented risks in terms of natural disasters, malicious acts, and systemic risk. Means to measure each were reviewed, along with methods commonly used to manage such risks. Natural disasters are a macro issue – with measurement undertaken by NGOs and governments. Malicious acts are also more a governmental/police matter. Systemic risks are of interest at both the macro and micro levels. Companies and firms have a number of resources to measure the risks they face and to manage them. The core of these resources would be their data and data management utilization strategies. These would include data collection, storing, analyzing, and information and knowledge generation. This book presents methodological tools and demonstration of several business cases implemented with freely available analytics software, i.e. R statistical programming language and R Studio.

Chapter 2 will give a quick introduction to R coding that we will use in the rest of the book. We continue with a discussion of enterprise risk management to include demonstration of basic data mining modeling of a credit card fraud dataset using R in Chapter 3. Chapter 4 will demonstrate association rule modeling in R with a supply chain management dataset. Simulation modeling can be done using R, but only by accomplishing each mathematical task and putting together a simulation model from scratch, which is quite cumbersome. Chapter 5 will discuss simulation modeling of enterprise risk management using Crystal Ball software. Chapter 6 will present forecasting modeling in R on the time series for S&P 500 data. This is followed by classification modeling in R demonstrated on China Credit Risk data in Chapter 7. Chapter 8 considers dataset balancing using SMOTE in R on a credit card fraud dataset. Chapter 9 will look at modeling categorical data using an HR Turnover dataset.

References

Abramov IL, Al-Zaidi ZAK (2023) Assessing the performance of construction companies in Iraq in terms of managing risk factors, *AIP Conference Proceedings*, 2936 https://doi.org/10.1063/5.0177948.

Araz OM, Choi TM, Olson DL, Salman FS (2020) Role of analytics for operational risk management in the era of big data. *Decision Sciences*, 51: 1320–1346.

Feldmeyer D, Birkmann J, McMillan JM, Stringer L, Leal Filho W, Djalante R, Pinho PF, Liwenga E (2021) Global vulnerability hotspots: Differences and agreement between international indicator-based assessments, *Climatic Change* 169(12): 11–32.

Forrest S, Milliken C (2018) Building resilience to disaster: From advice to action, *European Review* 27(1): 17–26.

Hochrainer-Stigler S, Colon C, Boza G, Brännström A, Linnerooth-Bayer J, Pflug G, Poledna S, Rovenskaya E, Dieckmann U (2020) Measuring, modeling, and managing systemic risk: The missing aspect of human agency, *Journal of Risk Research* 23(1): 1301–1317.

National Research Council (2012) *Disaster Resilience: A National Imperative*. Washington: The National Academies Press.

Olson DL (2015) Systems perspective of the primary aluminum supply chain: Unintended consequences of participant policies, *Human Systems Management* 34: 237–248.

Olson DL (2017) The real estate crash of 2007–8 as a systemic failure, *Human Systems Management* 35: 267–277.

2

R Programming Language and RStudio

R is a programming language and software environment highly suitable for statistical computing purposes. Specifically, it facilitates data manipulation and management, computation, and visualization. It is supported by many operating systems (Linux, Mac OS X, Windows) and available for free at www.r-project.org. R has become very popular recently and it is frequently used in various disciplines including bioinformatics, statistics, econometrics, psychology, machine learning, and statistical computing. Its use in many business applications is also developing rapidly, especially in business analytics, e.g., risk analytics, financial modeling, marketing, predictive analytics, auditing among many.

2.1 Loading R and RStudio

Once R is downloaded, the following graphic user interface (GUI) in Figure 2.1 would be available for the users. While it seems practical to execute simple operations in the console, it becomes more challenging with complex operations. Therefore, RStudio IDE can be used as a convenient editor, and that is what we use in this book as well. More information on RStudio can be found at the link www.rstudio.com and RStudio can be downloaded for free here: https://rstudio.com/products/rstudio/download/.

With RStudio IDE, instead of typing commands into R Console, you can generate commands by an editor and then send them to the R window by executing (Run) your code, and later you can modify them and send again. A sample view of RStudio is given in Figure 2.2, which displays the R script file on the upper left panel, the console on the lower left panel, the global environment with the data objects created during the work session on the upper right panel and finally the files, plots, packages, and help tools on the right bottom panel. RStudio provides flexibility so that the locations of these windows can be changed and moved around based on the users' will.

In this chapter, basic operations in R will be introduced and demonstrated in RStudio environment. As several different packages will be needed in the

DOI: 10.1201/9781003562399-2

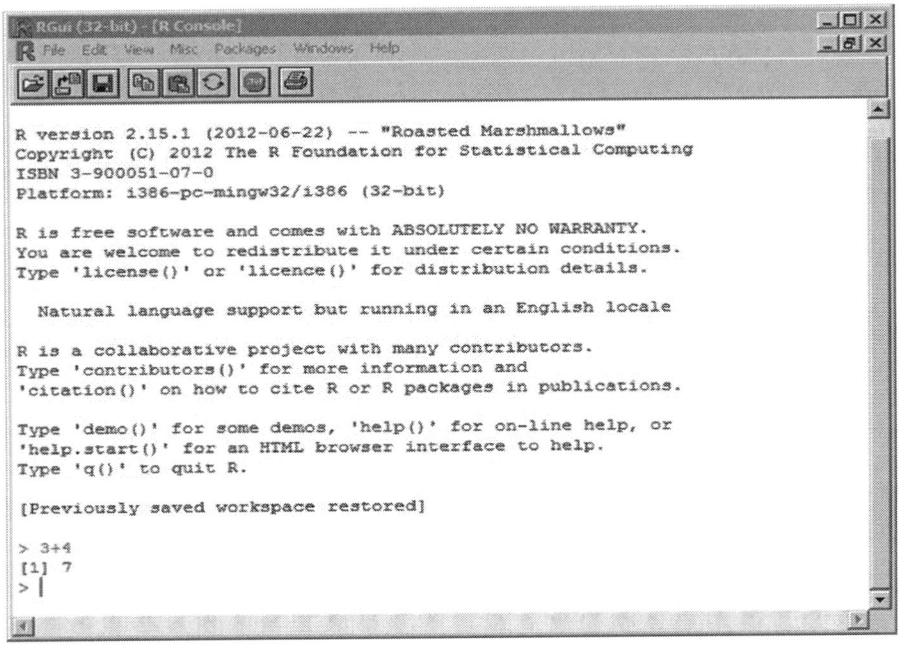

FIGURE 2.1
RGUI and RStudio.

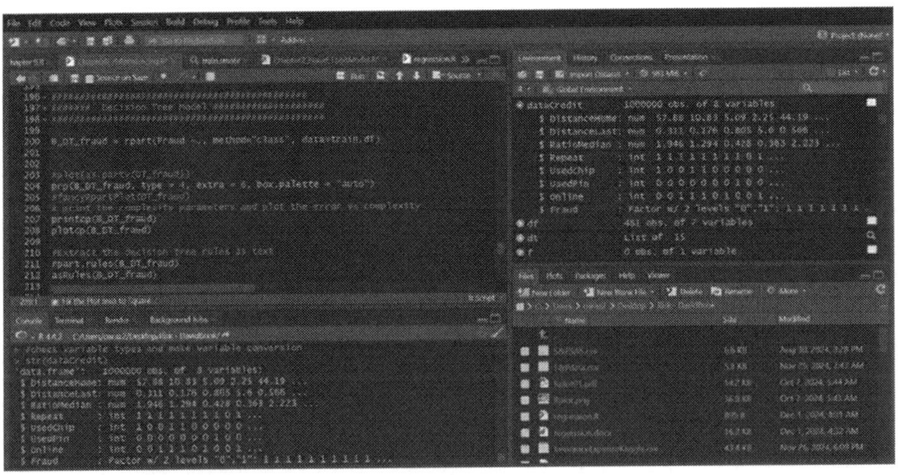

FIGURE 2.2
A view of RStudio IDE.

book for different modeling methods, we also demonstrate simple installing and loading operations for packages before a function from these packages can be called and utilized in a work session.

2.2 Data Frames and Reading Data Files

In this book, we generally use structured data files. After data cleaning and variable transformations, we read .csv files and load them as dataframes in RStudio. For example, CreditCardFraud.csv is one of the files used in this book. We would form a dataframe as a data object by reading this file as following after setting the working directory to the right folder where the file is stored:

```
data<-    as.data.frame(read.csv("CreditCardFraud.csv    ",
header=TRUE,sep=",")) .
Another way would be choosing the file of interest with
  file.choose() function in R.
data<-   as.data.frame(read.csv(file.choose(),       header=
  TRUE,sep=","))
```

To check the structure of the data object and the variables types, str() can be used, then necessary variables transformations can be identified as seen below.

```
data <- as.data.frame(read.csv("CreditCardFraud.csv"))
#check variable types and make variable conversion
str(data)
```

Figure 2.3 shows the result of these R commands.

For example, for classification purposes, we will need to recode the Fraud variable into a categorical variable, thus we can do that with the following script using as.factor() conversion. In R, as.numeric(), as.character() can also be used for necessary vector or string conversion.

```
data$Fraud= as.factor(data$Fraud)
```

Figure 2.4 displays the result of this R operation.

```
str(data)
'data.frame':	1000000 obs. of  8 variables:
 $ DistanceHome: num  57.88 10.83 5.09 2.25 44.19 ...
 $ DistanceLast: num  0.311 0.176 0.805 5.6 0.566 ...
 $ RatioMedian : num  1.946 1.294 0.428 0.363 2.223 ...
 $ Repeat      : int  1 1 1 1 1 1 1 1 0 1 ...
 $ UsedChip    : int  1 0 0 1 1 0 0 0 0 0 ...
 $ UsedPin     : int  0 0 0 0 0 0 0 1 0 0 ...
 $ Online      : int  0 0 1 1 1 0 1 0 0 1 ...
 $ Fraud       : int  0 0 0 0 0 0 0 0 0 0 ...
```

FIGURE 2.3
R dataframe creation.

```
str(data)
'data.frame':	1000000 obs. of  8 variables:
 $ DistanceHome: num  57.88 10.83 5.09 2.25 44.19 ...
 $ DistanceLast: num  0.311 0.176 0.805 5.6 0.566 ...
 $ RatioMedian : num  1.946 1.294 0.428 0.363 2.223 ...
 $ Repeat      : int  1 1 1 1 1 1 1 1 0 1 ...
 $ UsedChip    : int  1 0 0 1 1 0 0 0 0 0 ...
 $ UsedPin     : int  0 0 0 0 0 0 0 1 0 0 ...
 $ Online      : int  0 0 1 1 1 0 1 0 0 1 ...
 $ Fraud       : Factor w/ 2 levels "0","1": 1 1 1 1 1 1 1 1 1 1 ...
```

FIGURE 2.4
R data transformation outcome.

2.3 Working with R Packages

There are several built-in functions and datasets stored in packages for different purposes, e.g. time series, statistical tests, and visualizations. When a package is loaded in R work session, then its contents would become available. Therefore, packages must be downloaded and installed, then loaded before usage which can be performed via library(package_name) function.

Installing packages should be done using install.packages ("package_name") statement, before the package is loaded. For example, in this book we will frequently use "caret" package which can be installed and loaded as follows.

```
install.packages("caret")
library(caret)
```

In this book some of the packages that will be used the most can be listed as "rpart", "rpart.plot", "party", "partykit", "caret", "pROC", "randomForest", "xgboost", and several others. These are mostly used for specific model development, performance evaluation, ROC curve plotting, and other visualization purposes.

R Manual is the official introductory document for R programming which can be a reference for beginners to learn how to work with arithmetic operations, functions, loops, defining and working with variables, vectors and matrices (Venables et al., 1997). You can access the document in the following link.

http://cran.r-project.org/doc/manuals/R-intro.pdf.

2.4 Random Data Partitioning

One of the common practices in data mining is data partitioning for training, testing, and validation of the models. While the ratios for these can vary a typical application, the classification tasks in the following chapters will be using 70–80% of the data for the training of models and 20–30% for testing the models. To be able to perform a random selection for this partitioning to avoid any data selection bias, we set a seed for random number generation using set.seed() and sample() functions in R, and define an index vector to form our training and testing data. An example is given below using the Chinese Credit Risk data and case that we will use for logistic regression later in the book.

```
# China Credit Risk Data Set
CCreditRisk = as.data.frame(read.csv("Chinacredit_risk_
  dataset.csv"))
#variable conversion-
CCreditRisk$loan_status=as.factor(CCreditRisk$loan_
  status)
#data partitioning
set.seed(123)
training.index = sample(c(1:dim(CCreditRisk)[1]),
  dim(CCreditRisk)[1]*0.8)
training.data = CCreditRisk[training.index,]
test.data = CCreditRisk[-training.index,]
```

In this example, 80% of the data is used for training and the remaining 20% is used for model testing. dim() is used to measure the size of the data, which reports the number of rows and columns; and number of rows is used in partitioning. Therefore dim(CCreditRisk)[1] is used for finding the number of rows in the dataset.

Reference

Venables, WN, Smith, DM and the R Core Team (1997) Notes on R: A Programming Environment for Data Analysis and Graphics version 4.4.3: 2025-02-28.

3

Risk Measures in Finance and Insurance

The property and casualty insurance industry has suffered from intense price competition, leading to a condition of low profitability. Insurance is needed to protect against flood risks, along with preventive actions. For example, production facilities, especially in the oil refining business, require insurance. Insurers are happy to collect premiums, but they also expend efforts to educate those they insure to reduce the likelihood of problems. Insurance is also required in some countries for automobile operators. Both commercial and personal lines of insurances have been commoditized. There was a new record for catastrophic events in 2020, that included 22 in the United States. The economic contraction created by the pandemic in 2020 hit insurance harder than most industries (Javanmardian et al., 2021). One of the risks, however, is that policyholders know more than insurers about their probability of having accidents or their preventive efforts. Policyholders who are at greater risk tend to purchase more insurance coverage. Another risk is moral hazard, when those who purchase insurance cease to take preventive action.

A common tool used in risk management is insurance. Of course, the insured pay premiums, so they need to conduct a cost/benefit analysis when obtaining insurance. Even insurance companies need to consider the risk-transfer process, calculating the cost of reinsurance and diversification. Most property and casualty underwriters are investing in data and analytics to improve their performance. Best-in-class performers have put distance between themselves and competitors by building advanced data and analytics underwriting capabilities. These tools enable leading insurance carriers to improve risk evaluation and enhance efficiency. Risk segmentation models are built based on rules accumulated over time, with some enhancements from machine learning.

The basic measure used in financial risk management is value at risk (VaR).

DOI: 10.1201/9781003562399-3

3.1 Measurement of Financial Risk

VaR describes the probability distribution for the value (earnings or losses) of an investment (firm, portfolio, etc.). The mean is a point estimate of a statistic, showing historical central tendency. VaR is also a point estimate but offset from the mean. It requires specification of a given probability level and then provides the point estimate of the return or better at the prescribed probability. For instance, Figure 3.1 gives the distribution of Conditional-Value-at-Risk (CvaR) and VaR for a statistical distribution of returns.

However, VaR has undesirable properties, especially for gain and loss data with non-elliptical distributions. It satisfies the well-accepted principle of diversification under normal distribution. However, it violates the fairly accepted subadditive rule, i.e., the portfolio VaR is not smaller than the sum of component VaR. The reason is that VaR only considers the extreme percentile of a gain/loss distribution without considering the magnitude of the loss. As a consequence, a variant of VaR, usually labeled as CVaR, has been used. In terms of computational issues, optimization of CVaR can be very simple, which is another reason for adoption of the CvaR measure. This pioneer work was initiated by Rockafellar and Uryasev (2002), where CVaR constraints in optimization problems can be formulated as linear constraints. CVaR represents a weighted average between the VaR and losses exceeding the VaR. CVaR is a risk assessment approach used to reduce the probability a portfolio will incur large losses assuming a specified confidence level. It is possible to maximize portfolio return subject to constraints including CVaR and other downside risk measures, both absolute and relative to a benchmark (market and liability based). Nested simulation and least-squares

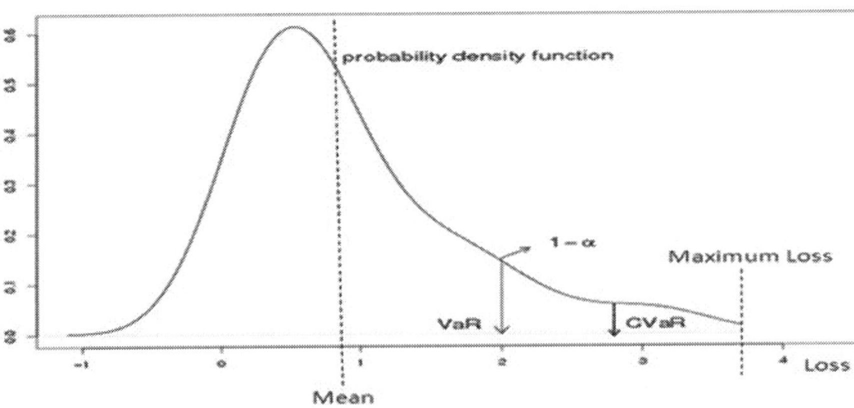

FIGURE 3.1
CVaR and VaR.

Monte Carlo simulation are effective means to obtain reliable estimates for VaR and CVar.

CVaR is a useful concept in terms of assessing probabilities of investment alternatives. It is a point estimator, like the mean (which could be viewed as the VaR for a probability of 0.5). It is only as valid as the assumptions made, which include the distributions used in the model and the parameter estimates. This is true of any simulation. However, CVaR provides a useful tool for financial investment. Monte Carlo simulation provides a flexible mechanism to measure it for any given assumption.

This chapter focuses on the tradeoffs between two approaches to optimize return subject to constraints on risk: CVaR and chance constraints. Microsoft Excel provides tools to optimize, but in order to generate more complex CVaR models, software tools such as MATLAB or R can be used. Simulation modeling also involves many complications.

3.2 Management of Financial Risk

The original concept of economic risk management was based on expected utility theory, recognizing individual risk attitudes. People tend to be risk averse when the magnitude of the risk is large, and risk neutral when that magnitude is small. Most initial studies of decision-making behavior used lotteries and other gambles, introducing a degree of complexity confusing some subjects.

Risks are traditionally viewed as a combination of probability and severity, although additional factors are also present in reality. Markowitz (1952) equated risk with variance, which would be controlled by diversification, measuring correlation across available investments and seeking efficient investment portfolios with respect to risk and return. Risk characteristics include uncertainties, dynamics, dependence, and complexities that motivate the use of various research tools. Financial risk management has developed such additional tools to include hedging, collars, VaR, and copulas.

3.2.1 Hedging

Hedging is another widely used tool for risk management. The idea is basic – sell a commodity at some future time at an agreed price to assure that the seller receives a minimal return. The downside of course is that the buyer is going to expect a premium. And if the price turns out to increase, the seller will lose out on that gain.

3.2.2 Investment Collars

A collar model applies an option strategy using puts and calls simultaneously to manage investment risk. A put option is purchased gaining the right to sell the underlying equity shares at a stated exercise price, providing downside protection. To offset the cost of the put, the investor sells a call option, where shares are sold at another exercise price. The collar refers to the range between the put and call options. Collars are intended to provide an alternative to diversification for risk management, inspired in part by the difficulties experienced in 2008 due to correlation across investments that were intended to be diversified but weren't. The benefit of the collar is to limit downside risk. The cost is that upside benefit is capped, as well as the cost of obtaining the options. There is also the risk of counterparty default on the put side.

3.2.3 Copulas

The basis of copulas is that bundles of investments with individual risks would be safer than each risk individually. The copula was a means to identify relative risk of a pool of risky assets compared to individual risky assets. Using a copula calculation assuming the normal distribution (a Gaussian copula), banks could bundle high-risk investments together in collateralized debt obligations (CDOs) that had much lower joint risk. CDOs are asset-backed securities that allow their issuer to use cash flows from different debt assets to back bonds that in turn could be sold to investors. The issuer could further market portions of these bonds associated with different levels of risk (tranches) which investors in turn would assume reflected different returns (assuming the higher risk yielded greater return while the pooled nature of the underlying assets reduced risks). This concept assumed stable correlations over time. The flaw in this logic was exposed in the 2008 real estate market.

3.3 Credit Card Fraud Data

Digital payments are evolving as are cyber criminals. According to the Data Breach Index, more than 5 million records are being stolen on a daily basis, a concerning statistic that shows fraud is still very common both for Card-Present and Card-Not Present payment types. We use the Credit Card Fraud data available at the Kaggle.com website at Dataset available from authors. This dataset contains 1,000,000 observations of credit card clients. Table 3.1 gives macro data for the Credit Card Fraud dataset.

TABLE 3.1

Credit Card Fraud Dataset Macro Data

Variable	Description	Type	Min	Max
distance_from_home	home to transaction point	number	0.0048	10632
distance_from_last_transaction	to last transaction	number	0.0001	11851
ratio_to_median_purchase_price	ratio	number	0.0044	267.8
repeat_retailer		binary		
used_chip	credit card chip used	binary		
used_pin_number		binary		
online_order		binary		
fraud	transaction is fraudulent	binary		

TABLE 3.2

Credit Card Fraud Dataset Initial Observations

distance_ home	distance_ from_last	ratio_to_ median	repeat_ retailer	used_ chip	used_pin_ number	online_ order	fraud
57.88	0.31	1.946	1	1	0	0	0
10.83	0.18	1.294	1	0	0	0	0
5.09	0.81	0.428	1	0	0	1	0
2.25	5.60	0.363	1	1	0	1	0
44.19	0.57	2.223	1	1	0	1	0
5.59	13.26	0.065	1	0	0	0	0
3.72	0.96	0.278	1	0	0	1	0
4.85	0.32	1.273	1	0	1	0	0
0.88	2.50	1.517	0	0	0	0	0
8.84	2.97	2.362	1	0	0	1	0
14.26	0.16	1.136	1	1	0	1	0
13.59	0.24	1.370	1	1	0	1	0
765.28	0.37	0.551	1	1	0	0	0
2.13	56.37	6.359	1	0	0	1	1
13.96	0.27	2.799	1	0	0	1	0

Table 3.2 displays the first fifteen observations from the dataset.

Table 3.3 displays counts for the binary data items in the Credit Card Fraud dataset.

3.3.1 Logistic Regression Model

In order to predict fraudulent transactions from this dataset, we first train a logistic regression model. Initially we explore the data in R. Initial data information from R can be obtained with the syntax given below after setting the working directory to the right folder in your machine and reading the data:

TABLE 3.3

Data on Credit Card Fraud Dataset Binary Variables

	Count	Fraud	RatioFraud	RatioNot
Total	1000000	87403	0.087	0.913
Repeat	881536	76925	0.087	0.913
Used Chip	350379	22410	0.064	0.936
Used Pin	100608	273	0.003	0.997
Online	650552	82711	0.127	0.873

```
##READ THE DATA FILE ###
dataCredit <- read.csv("CreditCardFraud.csv")
#check variable types and make variable conversion
```

This yields the output shown below:

```
str(dataCredit)

## 'data.frame':    1000000 obs. of  8 variables:
## $ DistanceHome: num  57.88 10.83 5.09 2.25 44.19 ...
## $ DistanceLast: num  0.311 0.176 0.805 5.6 0.566 ...
## $ RatioMedian : num  1.946 1.294 0.428 0.363 2.223 ...
## $ Repeat      : int  1 1 1 1 1 1 1 1 0 1 ...
## $ UsedChip    : int  1 0 0 1 1 0 0 0 0 0 ...
## $ UsedPin     : int  0 0 0 0 0 0 0 1 0 0 ...
## $ Online      : int  0 0 1 1 1 0 1 0 0 1 ...
## $ Fraud       : int  0 0 0 0 0 0 0 0 0 0 ...

summary (dataCredit)

## DistanceHome        DistanceLast         RatioMedian        Repeat
## Min.   :    0.005   Min.   :    0.000   Min.   :  0.0044   Min.   :0.0000
## 1st Qu.:    3.878   1st Qu.:    0.297   1st Qu.:  0.4757   1st Qu.:1.0000
## Median :    9.968   Median :    0.999   Median :  0.9977   Median :1.0000
## Mean   :   26.629   Mean   :    5.037   Mean   :  1.8242   Mean   :0.8815
## 3rd Qu.:   25.744   3rd Qu.:    3.356   3rd Qu.:  2.0964   3rd Qu.:1.0000
## Max.   :10632.724   Max.   : 11851.105  Max.   :267.8029   Max.   :1.0000
## UsedChip          UsedPin           Online            Fraud
## Min.   :0.0000   Min.   :0.0000   Min.   :0.0000   Min.   :0.0000
## 1st Qu.:0.0000   1st Qu.:0.0000   1st Qu.:0.0000   1st Qu.:0.0000
## Median :0.0000   Median :0.0000   Median :1.0000   Median :0.0000
## Mean   :0.3504   Mean   :0.1006   Mean   :0.6506   Mean   :0.0874
## 3rd Qu.:1.0000   3rd Qu.:0.0000   3rd Qu.:1.0000   3rd Qu.:0.0000
## Max.   :1.0000   Max.   :1.0000   Max.   :1.0000   Max.   :1.0000
```

TABLE 3.4

Credit Card Data Correlation Matrix

	distance_ home	distance_ from_last	ratio_to_ median	repeat_ retailer	used_ chip	used_ pin	online
distance_home	1.000						
distance_from_last	0.000	1.000					
ratio_to_median	-0.001	0.001	1.000				
repeat_retailer	0.143	-0.001	0.001	1.000			
used_chip	-0.001	0.002	0.001	-0.001	1.000		
used_pin	-0.002	-0.001	0.001	0.000	-0.001	1.000	
Online	-0.001	0.000	0.000	-0.001	0.000	0.000	1.000
Fraud	0.188	0.092	0.462	-0.001	-0.061	-0.100	0.192

Table 3.4 gives the correlation matrix of the variables in the dataset.

In order to train and test the logistic regression model, the dataset was partitioned in R using 90 percent for training and reserving 10 percent for testing. A seed is used to randomly partition the data into these two sets.

```
#data partitioning 90 % training and 10% test
set.seed(123)
train.index <- sample(c(1:dim(dataCredit)[1]),
dim(dataCredit)[1]*0.9)
train.df <- dataCredit[train.index, ]
valid.df <- dataCredit[-train.index, ]
```

This was followed by running a logistic regression model:

```
# run Logistic regression
# use glm() (general linear model) with family = "bino-
mial" to fit a Logistic
# regression.
fraud_model <- glm(Fraud ~ ., data = train.df,
family = "binomial")
```

This yielded a warning as the presence of binary variables can warp model assumptions (specifically the assumption of normality, but that really isn't an issue in most cases).

```
options(scipen=999)
summary(fraud_model)
```

Summary of the logistic regression output is displayed below:

```
##
## Call:
## glm(formula = Fraud ~ ., family = "binomial", data = train.df)
##
## Coefficients:
##                  Estimate   Std. Error z value              Pr(>|z|)
## (Intercept)  -10.34305230  0.05202636 -198.80 <0.0000000000000002 ***
## DistanceHome   0.01533944  0.00009978  153.73 <0.0000000000000002 ***
## DistanceLast   0.02559344  0.00028724   89.10 <0.0000000000000002 ***
## RatioMedian    0.86158784  0.00338719  254.37 <0.0000000000000002 ***
## Repeat1       -0.62060515  0.01885719  -32.91 <0.0000000000000002 ***
## UsedChip1     -1.04601118  0.01458781  -71.70 <0.0000000000000002 ***
## UsedPin1     -13.45354366  0.18656459  -72.11 <0.0000000000000002 ***
## Online1        6.63118845  0.04427920  149.76 <0.0000000000000002 ***
## ---
## Signif. codes:  0 '***' 0.001 '**' 0.01 '*' 0.05 '.' 0.1 ' ' 1
##
## (Dispersion parameter for binomial family taken to be 1)
##
##     Null deviance: 415945  on 699999  degrees of freedom
## Residual deviance: 188528  on 699992  degrees of freedom
## AIC: 188544
##
## Number of Fisher Scoring iterations: 10
```

The results of the logistic regression show all variables to be highly significant. The Akaiki information criterion (AIC gives a measure of model fit, with lower scores being better). The model displayed above can be used to forecast, given values for the seven independent variables. The next output shows the application of predictions with this model in R applied to ten records:

```
# use predict() with type = "response" to compute
predicted probabilities.

fraud.pred <- predict(fraud_model, valid.df[, -8],
type = "response")
```

```
# first 10 actual and predicted records
data.frame(actual = valid.df$Fraud[1:10],
predicted = ifelse(fraud.pred[1:10]>0.5, 1,0))

##    actual predicted
## 5       0         0
## 8       0         0
## 11      0         0
## 22      0         0
## 36      1         0
## 37      1         1
## 39      0         0
## 40      0         0
## 58      0         0
## 95      0         0
```

R Model Fit can be obtained with the syntax:

```
# Form a table from the output-- CONFUSION MATRIX ----
#cm=table(logit.reg.pred>0.5, valid.df$Personal.Loan)
#cm
#write.csv(cm,"cm.csv")

#install caret and use confusionMatrix()

### PLOTING AUCs#####
install.packages(c("rattle",    "rpart",    "rpart.plot",
"party", "partykit", "caret", "pROC"), repos = "http://
cran.us.r-project.org")
# call packages and load

library(caret)

## Loading required package: ggplot2

## Warning: package 'ggplot2' was built under R version 4.3.3

## Loading required package: lattice

fraud.pred= factor(ifelse (fraud.pred >0.5, 1, 0))
confusionMatrix(data=fraud.pred,   reference   =   valid.
df$Fraud, mode = "everything", positive = "1")
```

This yields model fit statistics displayed here, beginning with the confusion matrix:

```
## Confusion Matrix and Statistics
##
##           Reference
## Prediction     0     1
##          0 90792  3343
##          1   624  5241
##
##                Accuracy : 0.9603
##                  95% CI : (0.9591, 0.9615)
##     No Information Rate : 0.9142
##     P-Value [Acc > NIR] : < 0.00000000000000022
##
##                   Kappa : 0.7049
##
##  Mcnemar's Test P-Value : < 0.00000000000000022
##
##             Sensitivity : 0.61055
##             Specificity : 0.99317
##          Pos Pred Value : 0.89361
##          Neg Pred Value : 0.96449
##               Precision : 0.89361
##                  Recall : 0.61055
##                      F1 : 0.72545
##              Prevalence : 0.08584
##          Detection Rate : 0.05241
##    Detection Prevalence : 0.05865
##       Balanced Accuracy : 0.80186
##
##        'Positive' Class : 1
##
```

The area under the curve for this model can be obtained with the syntax:

```
library(rpart)
library(party)
library(caret)
library(pROC)
## Type 'citation("pROC")' for a citation.
```

```
##
## Attaching package: 'pROC'

## The following objects are masked from 'package:stats':
##
##     cov, smooth, var

library(rpart.plot)
## Warning: package 'rpart.plot' was built under R
version 4.3.3
library(rattle)
fraud.pred <- predict(fraud_model, valid.df[, -8],
type = "response")

#test_roc = roc(valid.df$Fraud ~ fraud.pred, plot = TRUE,
print.auc = TRUE)
###Fix the Plot area to Square####
par(pty="s")
test_roc = roc(valid.df$Fraud ~ fraud.pred, plot = TRUE,
print.auc = TRUE, legacy.axes=TRUE, percent=TRUE, xlab=
"False Positive Rate", ylab="True Positive Rate", lwd=4)
## Setting levels: control = 0, case = 1
## Setting direction: controls < cases
```

Figure 3.2 shows the ROC Plot for this model with an area under the curve of 0.968.

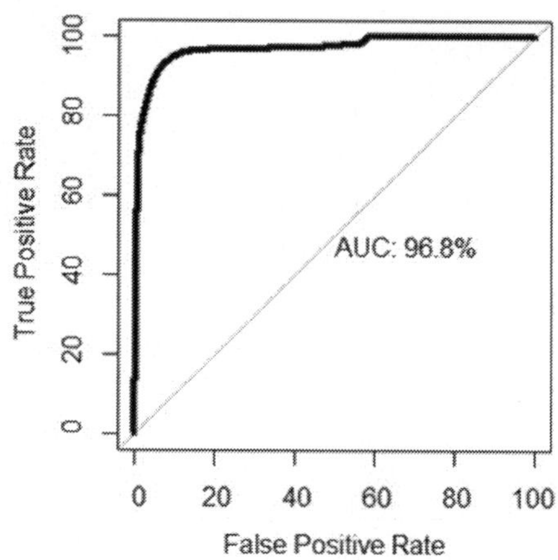

FIGURE 3.2
ROC Plot for Logistic Regression Model.

```
###Return back to Maximum Plot Area###
par(pty="m")
```

3.3.2 Decision Tree Model

A decision tree classification model can also be utilized for the same purpose. More details on decision tree models are given in Chapter 7. The R syntax to generate a decision tree using the training dataset is:

```
#Decision Tree Model
DT_fraud = rpart(Fraud ~., method="class", data=train.df)

#plot(as.party(DT_fraud))
prp(DT_fraud, type = 4, extra = 6, box.palette = "auto")
```

Figure 3.3 displays the resulting decision tree.

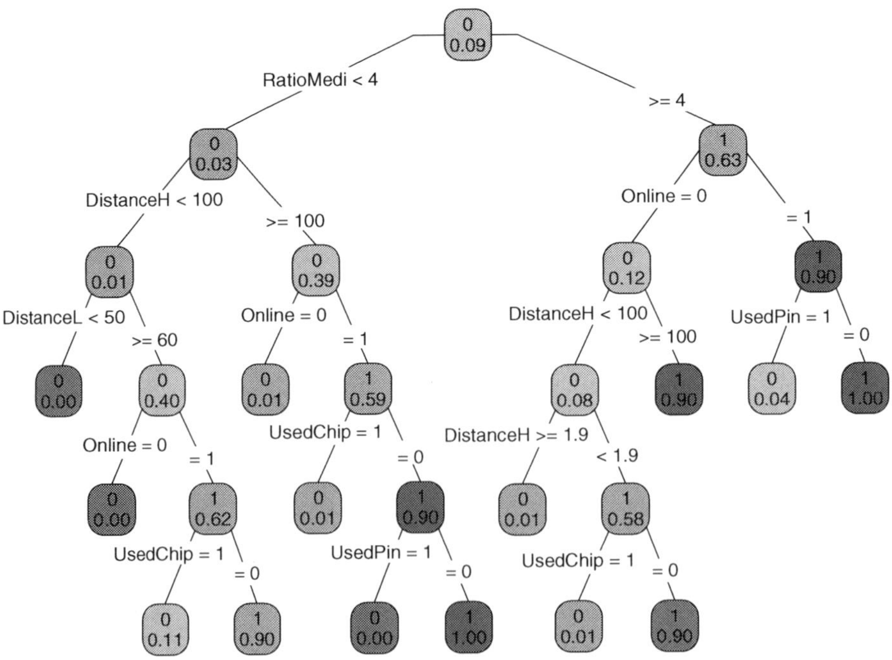

FIGURE 3.3
R Decision Tree.

Details of the decision tree model are obtained with the following syntax:

```
#fancyRpartPlot(DT_fraud)
# print the complexity parameters and plot the error vs
complexity
printcp(DT_fraud)

##
## Classification tree:
## rpart(formula = Fraud ~ ., data = train.df,
method = "class")
##
## Variables actually used in tree construction:
## [1] DistanceHome DistanceLast Online       RatioMedian
UsedChip
## [6] UsedPin
##
## Root node error: 78819/900000 = 0.087577
##
## n= 900000
##
##          CP nsplit rel error   xerror      xstd
## 1 0.309139      0  1.000000 1.000000 0.00340238
## 2 0.070833      2  0.381723 0.381723 0.00216359
## 3 0.057816      3  0.310890 0.310890 0.00195881
## 4 0.021137      6  0.137441 0.137454 0.00131260
## 5 0.016379      7  0.116304 0.116317 0.00120860
## 6 0.016058      8  0.099925 0.103681 0.00114170
## 7 0.012364     11  0.051751 0.051790 0.00080876
## 8 0.010000     13  0.027024 0.027062 0.00058526
```

The syntax to obtain a plot of relative error by tree size is:

```
plotcp(DT_fraud)
```

This yields the plot shown in Figure 3.4.

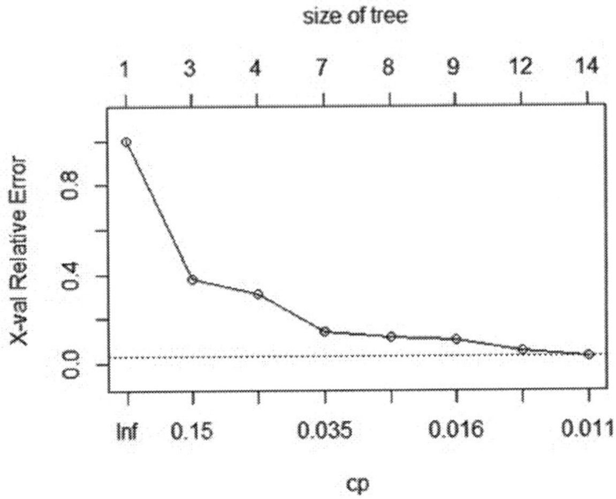

FIGURE 3.4
Relative Error by Tree Size.

The following syntax yields a list of rules that are displayed in the tree model:

```
#Extract the decision tree rules as text

asRules(DT_fraud)

##
## Rule number: 15 [Fraud=1 cover=54315 (6%) prob=1.00]
##   RatioMedian>=4
##   Online>=0.5
##   UsedPin< 0.5
##
## Rule number: 47 [Fraud=1 cover=15337 (2%) prob=1.00]
##   RatioMedian< 4
##   DistanceHome>=100
##   Online>=0.5
##   UsedChip< 0.5
##   UsedPin< 0.5
##
## Rule number: 13 [Fraud=1 cover=1597 (0%) prob=0.90]
##   RatioMedian>=4
##   Online< 0.5
##   DistanceHome>=100
##
## Rule number: 39 [Fraud=1 cover=4783 (1%) prob=0.90]
```

```
##     RatioMedian< 4
##     DistanceHome< 100
##     DistanceLast>=50.03
##     Online>=0.5
##     UsedChip< 0.5
##
##  Rule number: 51 [Fraud=1 cover=2459 (0%) prob=0.90]
##     RatioMedian>=4
##     Online< 0.5
##     DistanceHome< 100
##     DistanceHome< 1.9
##     UsedChip< 0.5
##
##  Rule number: 38 [Fraud=0 cover=2646 (0%) prob=0.11]
##     RatioMedian< 4
##     DistanceHome< 100
##     DistanceLast>=50.03
##     Online>=0.5
##     UsedChip>=0.5
##
##  Rule number: 14 [Fraud=0 cover=6087 (1%) prob=0.04]
##     RatioMedian>=4
##     Online>=0.5
##     UsedPin>=0.5
##
##  Rule number: 10 [Fraud=0 cover=14062 (2%) prob=0.01]
##     RatioMedian< 4
##     DistanceHome>=100
##     Online< 0.5
##
##  Rule number: 24 [Fraud=0 cover=27166 (3%) prob=0.01]
##     RatioMedian>=4
##     Online< 0.5
##     DistanceHome< 100
##     DistanceHome>=1.9
##
##  Rule number: 22 [Fraud=0 cover=9141 (1%) prob=0.01]
##     RatioMedian< 4
##     DistanceHome>=100
##     Online>=0.5
##     UsedChip>=0.5
##
```

```
##  Rule number: 50 [Fraud=0 cover=1363 (0%) prob=0.01]
##    RatioMedian>=4
##    Online< 0.5
##    DistanceHome< 100
##    DistanceHome< 1.9
##    UsedChip>=0.5
##
##  Rule number: 8 [Fraud=0 cover=755425 (84%) prob=0.00]
##    RatioMedian< 4
##    DistanceHome< 100
##    DistanceLast< 50.03
##
##  Rule number: 46 [Fraud=0 cover=1666 (0%) prob=0.00]
##    RatioMedian< 4
##    DistanceHome>=100
##    Online>=0.5
##    UsedChip< 0.5
##    UsedPin>=0.5
##
##  Rule number: 18 [Fraud=0 cover=3953 (0%) prob=0.00]
##    RatioMedian< 4
##    DistanceHome< 100
##    DistanceLast>=50.03
##    Online< 0.5
```

These rules are useful in explaining results. Of the fourteen rules obtained, five have higher probabilities of fraud (the first five listed). Rule 15 consists of relatively higher priced purchases online without use of a pin. The other higher level purchase rules are rule 13 (not online, far from home) and rule 51 (not online, very close to home, no chip used). Two rules predicted fraud for less expensive purchases. Rule 47 involved purchases away from home, online, with no chip and no pin. Rule 39 involved purchases close to home, but at a distance from the last purchase, online, with no chip. The other 9 cases had low probabilities of fraud.

Confusion matrices provide a means to measure the relative accuracy of models. The following syntax gives the confusion matrix for the decision tree along with fit statistics:

```
pred_DT1 = predict(DT_fraud, valid.df, decision.values=
TRUE, type="class")
confusionMatrix(data=pred_DT1, reference = valid.df$Fraud,
mode = "prec_recall", positive = "1")
```

This results in the output shown:

```
## Confusion Matrix and Statistics
##
##           Reference
## Prediction     0     1
##          0 91310   130
##          1   106  8454
##
##                Accuracy : 0.9976
##                  95% CI : (0.9973, 0.9979)
##     No Information Rate : 0.9142
##     P-Value [Acc > NIR] : <0.0000000000000002
##
##                   Kappa : 0.9849
##
##  Mcnemar’s Test P-Value : 0.1343
##
##               Precision : 0.9876
##                  Recall : 0.9848
##                      F1 : 0.9862
##              Prevalence : 0.0858
##          Detection Rate : 0.0845
##    Detection Prevalence : 0.0856
##       Balanced Accuracy : 0.9918
##
##        ‘Positive’ Class : 1
##
```

The accuracy of this model was very strong, over 99 percent. To obtain the area under the curve R syntax is:

```
# ROC and AUC for DT1 model
test_prob2 = predict(DT_fraud,valid.df[,-8], decision.
values= TRUE, type=“prob”) [,2]
###Fix the Plot area to Square####
par(pty=“s”)
test_roc = roc(valid.df$Fraud ~ test_prob2, plot = TRUE,
print.auc = TRUE, legacy.axes=TRUE, percent=TRUE, xlab=
“False Positive Rate”, ylab=”True Positive Rate”, lwd=4)
```

```
## Setting levels: control = 0, case = 1
## Setting direction: controls < cases
```

This yields an area under the curve of 0.992, a very good fit on the test data.

3.4 Conclusions

This chapter discussed financial risk management. Conventional finance tools were briefly reviewed, with focus on data mining tools available in R. A dataset on credit card fraud from the Kaggle.com website was used. Logistic regression was demonstrated as well as decision tree modeling. For this data, the decision tree modeled 14 rules with a very strong fit on the training data.

Future chapters will go deeper into the process of data mining modeling in R.

References

Javanmardian K, Ramezani Sm, Srivastava Am Talischi C (2021) How data and analytics are redefining excellence in P&C underwriting, *McKinsey White Paper*. McKinsey.

Markowitz H (1952) Portfolio selection, *The Journal of Finance* 17(1): 77–91.

Rockafellar R, Uryasev S (2002) Conditional value-at-risk for general loss distributions, *Journal of Banking and Finance* 26: 1443–1471.

4

Association Rule Modeling in Supply Chains

Supply chains are critical elements in global economies, as they have been for millennia. The silk road operated throughout civilized human history and continues today as a major Chinese initiative in globalization. Military organizations over that same period have focused on logistics, the field of getting the material needed for their organizations to move over vast distances (from Alexander the Great, the Roman legions, the Spanish colonization of South America, through modern era efforts such as World War II and more recent operations).

But militaries aren't the only organizations reliant upon supply chains. Lean manufacturing relies upon the ability to source and move materials quickly to manufacturing plants (Toyota is credited with pioneering efforts at reducing in-bound inventories). Other approaches include make-to-order manufacturing, reducing finished goods inventories. Major retail operators, such as Wal-Mart, today rely on supply chains linked with massive computer support. Amazon has revolutionized goods delivery relying upon computer support and a dense network of delivery systems to drastically cut order-to-delivery times.

Supply chains enable a great deal of consumer benefit offering access to goods from around the globe. The food supply chain is remarkable in enabling customers in Canada to access tropical fruits in a fresh state, customers in Denmark to obtain Argentinian beef, customers in Shanghai to obtain coffee fresh from Africa. But of course there are risks involved in supply chains; with the COVID-19 pandemic of 2020 the vulnerability of global supply chain operations has become more visible. Supply chains involve flows of material, information, and money. Risks are present in the technological, economic, financial, performance, and legal/regulatory areas.

The efficiency ideas obtained from just-in-time and make-to-order manufacturing depend upon reliable supply chains accessing goods from around the globe. Chinese manufacturing grew to become a major source of goods used throughout the world. The pandemic resulted in plant shutdowns that disrupted many supply chains. Further, the logistic networks clogged due to port shutdowns. There were waiting lines of cargo ships months long in California ports at one time. Many industries were blocked from obtaining needed materials. This was especially acute in cases such as parts made of

DOI: 10.1201/9781003562399-4

rare earths, often obtained only from China. Even within Europe and the United States, many plants shut down when COVID-19 appeared in their workforce.

This chapter will begin with a brief discussion of things that can go wrong in supply chain management. This will be followed by discussion of a risk management framework and risk management process. The focus of the chapter is on association rule modeling to support supply chain risk management.

4.1 Supply Chain Risk Measures

First, we will consider things that could go wrong. Supply chains around the world were hit with a massive pandemic in 2020. But of course, pandemics are not the only risks faced by supply chains. Wicaksana et al. (2022) viewed supply chain risks from multiple perspectives to include location. Supply-side risks include supplier downtown and quality, relationships with suppliers, insolvency, and discontinuance of spare parts. Manufacturing-side risks included failures in operations planning and mismatches between supply and demand. Demand-side risks included volume variance and problems with shipping. The flows considered were flow of money (to include exchange rate fluctuation), flow of goods (disruption of deliveries for various reasons), and flow of information (inadequate feedback loops within the supply chain as well as cyber security).

Wang et al. (2020) viewed three supply chains, each with their own set of specific risks and means to manage those risks. In the agricultural sector, production yield is uncertain, in part due to weather, climate, and hazards in storage, handling, and processing. Agricultural goods usually are perishable, creating additional risks. Methods to manage these risks include forward market (selling contracts prior to harvesting). Leasing arrangements also can be used to alleviate some risks. The energy sector differs in that most energy commodities are storable. There is also a mature financial market for energy commodities (as there are for agricultural products). Metals are important in many industries to include power, construction, and manufacturing. The greatest risk is price volatility. Financial hedging is frequently used to manage price risk. For general industrial commodities without financial derivative markets, risk mitigation needs to be managed through use of the spot market and using option/forward contracts. There is a need to forego apparent profit optimization to take into account the countervailing needs of risk management (usually you have to sacrifice some profit to reduce risk).

Table 4.1 lists some major supply chain risks.

TABLE 4.1

Supply Chain Risk Categories

Process Category	Sub-category	Risks
Supplier	Disruption	Supplier insolvency
		Production halt
		Industrial zone shutdown
	Quality	Quality degradation
	Outages	Pandemic
		Accidents
	Technology	Denial of service
		Ransomware
		Cyber attack
Demand	Customer	Economic downturn
	Competition	Inflation
		Labor disputes
		Customer payment
		Product obsolescence
Transportation (logistics)	Ground	Strikes
		Accidents
		Routing & scheduling failures
	Marine	Port congestion
		Port closure
		Piracy
	Aviation	Customs delay
		Airport closure
Socio-political	Political violence	Terrorism
		Civil unrest
		War
	Regulatory	Tax policies
		Corruption
		Government seizures
	Security	Disease outbreak
		Trade regulations
		Fire
Nature (environmental)	Weather	Hurricanes
		Hail
		Lightning
		Flood
		Climate change
	Geology	Volcano
		Tsunami

Sources: Adapted from DuHadway et al. (2019), Buchholz et al. (2022), and Olson and Wu (2023).

There are more specific risk categories for particular aspects of supply chains. Olson and Chae (2023) examined supply chain firm filings to identify reported risks. Table 4.2 displays examples for the field of transportation.

TABLE 4.2
Risk Categories in Transportation

Category	Trucking	Shipping	Air	Rail
Personnel	Recruitment Driver retention Unionization	Manning crew Captain	Furloughs Pilots Attendants Dispatchers Union	Injury
Fuel	Diesel	Gas Diesel Hydrocarbon	Jet fuel	Coal
Weather	Snow, Ice	Ice	Cancellations	
Risks	Terrorism Fraud Emissions	Leakage Pollution Vice	War Emissions Noise	Crime Clean-up War
Regulatory	Licensing Hours of service Taxation	Inspection Treaties Certification	FAA Warrants	NAFTA
Pandemic	COVID-19	COVID-19	COVID-19	COVID-19

Source: Extracted from Olson and Chae (2023).

That study of course found mention of COVID-19, which drastically impacted global supply chains, with manufacturing plants often shutting down and transportation links blocked by port closures, leading to excessively long lines of ships waiting to get into port for months. The value of alternative supplies became obvious, but it turned out that these alternatives have pandemic issues as well. Otherwise, each transportation mode had its own set of risks.

4.2 Supply Chain Risk Management Framework

Supply chains consist of a number of elements:

a. Vendor selection (sourcing)
b. Logistics (transportation of inputs to production facilities)
c. Demand management (advertising to generate customer demand)
d. Production (making or processing marketable products)
e. Retail channels (outlets to deliver product to customers)

Risks occur at each of these supply chain elements. We will first discuss a risk management process, followed by discussion of how association rule mining has been applied to various elements of supply chains.

4.2.1 Supply Chain Risk Management Process

A typical process applied to the supply chain risk management process includes steps if risk identification, risk assessment, risk avoidance, and risk mitigation. These steps provide a framework under which robust strategies can be designed to cope with the many risks involved in global supply chains.

4.2.2 Risk Identification

Risks in supply chains can include operational risks and disruptions. Operational risks involve inherent uncertainties for supply chain elements such as customer demand, supply, and cost. Disruption risks come from disasters (natural in the form of floods, hurricanes, etc.; man-made in the form of terrorist attacks or wars) and from economic crises (currency reevaluations, strikes, and shifting market prices). Most quantitative analyses and methods are focused on operational risks. Disruptions are more dramatic, less predictable, and thus are much more difficult to model and manage. Risk management planning and response for disruptions are usually qualitative.

Lean manufacturing gains efficiencies by reducing inventory stocks, which of course serves as a form of insurance against supply disruption. Media often blames lean systems for disrupted production. However, even when safety stocks were larger, there were still supply disruptions. The added risk due to lean safety stocks is probably more than offset by the gains in reduced inventory expense.

4.2.3 Risk Assessment

Theoretically, risk has been viewed as applying to those cases where odds are known (like casinos) and uncertainty to those cases where odds are not known (like investment in a new venture). Risk is a preferable basis for decision-making, but life often presents decision-makers with cases of uncertainty. The issue is further complicated in that perfectly rational decision-makers may have radically different approaches to risk. Research into cognitive psychology has found that managers are often insensitive to probability estimates of possible outcomes and tend to ignore possible events that they consider to be unlikely. Furthermore, managers tend to pay little attention to uncertainty involved with positive outcomes. They tend to focus on critical performance targets, which makes their response to risk contingent upon context. Some approaches to theoretical decision-making prefer objective treatment of risk through quantitative scientific measures following normative ideas of how humans should make decisions. Business involves high levels of uncertainty (data not available) and consideration of multiple (often conflicting) factors, making qualitative approaches based upon perceived managerial risk more appropriate.

Because accurate measures of factors such as probability are often lacking, robust strategies (more likely to enable effective response under a wide range of circumstances) are often attractive to risk managers. Strategies are efficient if they enable a firm to deal with operational risks efficiently regardless of major disruptions. Strategies are resilient if they enable a firm to keep operating despite major disruptions. Supply chain risk can arise from many sources, to include product availability, distance from sources, industry capacity, changes in technology and labor markets, financial instability, and management turnover, as well as those internal risk categories.

4.2.4 Risk Avoidance

The oldest form of risk avoidance is probably insurance, purchasing some level of financial security from an underwriter. This focuses on the financial aspects of risk, and is reactive, providing some recovery after a negative experience. Insurance focuses on financial risks. Other major risks include loss of customers due to supply change disruption.

Supply chain risks can be buffered by a variety of methods. Purchasing is usually assigned the responsibility of controlling costs and assuring continuity of supply. Buffers in the form of inventories exist to provide some risk reduction, at a cost of higher inventory holding cost. Traditional practice, relying upon extra inventory, multiple suppliers, expediting, and frequent supplier changes, suffers from high transaction costs, long purchase fulfillment cycle times, and expensive rush orders. Newer risk management approaches, drawing upon practices such as supply chain alliances, e-procurement, just-in-time delivery, increased coordination, and other techniques, provide more visibility in supply chain operations. There may be higher prices incurred for goods, and increased security issues, but supply chain risks are reduced.

4.2.5 Risk Mitigation

There are both strategic and tactical aspects involved in supply chain risk. Strategically, a network design of supply chain participants can enable better control of supply risks through activities such as finding alternative supply sources in times of emergency. Strategies such as product pricing and rollovers can control demand to a degree. Greater product variety can strategically protect against product risks. And systems providing greater information visibility across supply chain members can enable better coping with risks. Tactical decisions include supplier selection and order allocation (including contractual arrangements); demand control over time, markets, and products; product promotion; and information sharing, vendor managed inventory systems, and collaborative planning, forecasting, and replenishment.

4.3 Mitigation Strategies

Mitigation is how risks are treated. Some basic mitigation activities include expanding capacity where you have competitive advantage. In the inventory area, buffers, safety stocks, and redundant suppliers have benefit, although this counteracts lean initiatives. With respect to sources, redundant suppliers are useful, along with a policy of multiple sourcing and avoiding suppliers in countries with high geopolitical risk. Increasing flexibility can be accomplished by utilizing product differentiation and delaying resource commitment. Financially, hedging, insurance, and global dispersion are viable strategies. Every strategy balances some tradeoffs, such as safety stock and lean policies.

A brief listing of risk response strategies is:

- Risk avoidance
  - Risks are critical and could stop the business
  - Suitable if high probability and high impact
- Risk reduction
  - Suitable if high probability and low impact
  - Often for operational risks
  - Techniques: redundancy, more quality testing, training
- Risk transfer
  - Suitable when low probability and high impact
  - Examples include natural disaster and terrorist attack
  - Techniques: insurance, contracts
- Risk acceptance
  - Suitable when low probability and low impact
  - Cost of reducing risk is much higher than expected impact
  - Techniques: contingency plans
- Ignore risk
  - Suitable when low probability and low impact
- Exploit risk
  - Risk impact might offer competitive advantage if organization develops means to cope economically

Olson and Wu (2023) reviewed some past studies of risk mitigation strategies in supply chain management (see Table 4.3).

TABLE 4.3
Risk Mitigation Strategies

Risk Category	Mitigation	Elements	Robust Strategy
Supply	Multiple suppliers Flexible contracts Aid supplier improvement	Redundancy	Strategic stock Economic incentives to suppliers
Processing	Flexible production Modularization Time product introduction	Flexibility	Postponement Make-or-buy Flexible transport
Demand	Shift demand over time Customize service	Agility	Influence customer selection Dynamic assortment planning
Network	Coordinate decisions Information sharing	Collaboration	

4.4 Information Technology Support

Identifying and assessing risks in supply chains is key to supply chain risk management. Computer software, to include artificial intelligence, is being used to provide rapid and adaptive risk management decision-making. Baryannis et al. (2019) reviewed application of artificial intelligence to supply chain risk management through text mining. They defined artificial intelligence in that context to provide systems that were capable of autonomously selecting courses of action in partially unknown supply chain environments. The tools they identified being used included stochastic programming, fuzzy programming, optimization, machine learning and big data, and agent-based methods. They concluded that reports focus more on structured and developed mathematical programming models, while the future impact is probably more to be found in machine learning, automated reasoning, and multi-agent systems.

4.4.1 Agricultural Supply Chain Risks

Wang (2024) looked at the information flow in agricultural product supply chains in China. Information on demand, funding, and logistics are coordinated using e-commerce platforms to connect producers, distributors, and retailers. External factors included natural hazards and environmental pollution. Internal factors for the production chain included agricultural inputs as well as plant and animal diseases. In the transportation network, information on storage, additive abuse, and processing hygiene were critical. On the distribution side, information on testing and environmental hygiene were needed. Wang (2024) surveyed 80 employees of an agricultural supply chain firm and 20 more of relevant professionals. The highest ranked risk indicators were transport and storage techniques, misuse of additives, substandard processing conditions, and substandard marketing hygiene. The primary point is that e-commerce platforms provide means for better and more rapid flow of information to better enable control and mitigation of risks in agricultural supply chains.

4.4.2 Blockchain Technology

Blockchains provide access to computer systems that are accurate, reliable, visible to authorized users, and incorruptible (in the sense that they are unchangeable). They do involve high levels of electrical use and cloud computing assets. Chowdhury et al. (2022) examined adoption of blockchain technology in UK supply chain risk management, proposing that this technology provides an opportunity to enhance supply chain risk management – especially in the context of emergencies, crises, disasters, and pandemics. That study surveyed 116 operations managers in the UK during the COVID-19 pandemic and applied structural equation modeling to examine benefits of blockchain use.

That study's literature review noted use of blockchain technology by major supply chain operations such as Wal-Mart in the food retail area, providing consumers to trace product origins while streamlining the restocking process. Other applications reported included Maersk managed international goods movements by using blockchains to track cargo movement, enabling reduction of waiting time through eliminating paper transactions. British Airways obtained better information accuracy across multiple platforms about flight information, linking Websites, airport gate monitors, and mobile apps. DHL reduced tampering and counterfeiting of drugs by blockchain provision of transparency, traceability, and trackability. Fedex stored data gathered from sensors and customers to resolve issues faster using permissioned ledgers across their transportation supply chain.

Association rule mining techniques are developed and used in enterprise risk management. We can also identify published applications of association rule modeling to various elements of the supply chain process.

Tsang et al. (2022) applied fuzzy association rule mining (along with genetic algorithm application to select options) to a new product development case involving a Chinese furniture manufacture producing curtain products. Input data about customers, engineering capabilities, market uncertainties, and product features were fed into a fuzzy association rule mining model to identify frequent itemsets to reduce problems in product design. This process aided in prototype development and market positioning. Association rules contributed by identifying the most likely levels of input relationships to product design features.

Jain et al. (2007) applied fuzzy association rule mining to supplier selection considering the criteria of cost, quality, service level, cycle time, and organizational features (to include reliability, dependability, and ability to cope with risks). Association rules were generated by converting the data to boolean form and identifying vendor evaluation features that made sense to consider together. An example rule was that IF cost were evaluated THEN quality should be evaluated. They demonstrated their model considering five suppliers.

In logistics, Ting et al. (2014) applied association rule mining to a food quality case involving a Hong Kong red wine company that had been experiencing quality and safety issues in their global supply chain (importing wine from Italy to Hong Kong). Rules were generated representing the relationship between logical order flows to source relationships (such as transportation mode, product type, and delivery progress). An example rule would be IF Product A was shipped on date B from factory C to retailer D THEN quality was X. In the case, the association rule model led to improved quality and reduced inspection cost.

In the production operations area, Lee et al. (2014) also applied fuzzy association rule mining to a real case study involving a garment manufacturer. Their model considered six inputs and two outputs (major or minor defects per garment). An example rule obtained was:

IF production time before delivery is short

AND sewing distance is long

AND the number of trims for attachment per garment is large

AND the average number of minor defects per garment is small

THEN the speed of the sewing machine should be adjusted to high

AND the speed of the finishing machine should be adjusted to high.

In the retail channel area, Ellis et al. (2018) applied cluster analysis along with association rule mining for three major retailers who had experienced inefficiencies in their use of RFID. RFID tag parameters were analyzed to identify better reading rates. A dataset of over 45,000 observations linking RFID tag lab test data with real world performance (good tag; bad tag) was used, generating rules identifying good or bad tags for given input characteristics. An example rule might be IF Antenna X was low THEN good tag. The association rule analysis identified patterns of purchase behavior.

4.5 Association Rules

Association rules seek to identify combinations of things that frequently occur together (**affinity analysis**). Association rules apply a form of machine learning, the most common of which is the apriori algorithm. Structured data is saved in fixed fields in databases and traditionally focused on quantitative data that could be analyzed by classical statistics. Data comes in many forms, including text, video, voice, image, and multimedia. Association rules are typically applied to data in non-quantitative form.

Association rules can provide information that can be useful in a number of ways. In the field of healthcare, many applications can be seen with clinical and population health data, for example they have been used to:

- Assess patterns of multimorbidity between hypertension, cardiovascular disease, and affective disorders in elderly patients;
- Analyze electronic health record databases to identify comorbidities;
- Monitor common diseases in particular areas and age groups to identify shortages of doctors;
- Identify patient problems when monitoring real-time sensor data to analyze abnormality in physiological conditions.

Association rules are of the IF-THEN type. Unfortunately, the more variables (products), the more combinations, which makes algorithms take a lot longer computational time and generate a lot more output to interpret. Some key concepts in association rule mining include coverage (**support**), which is the number of instances occurring in the dataset, and accuracy (**confidence**), correct prediction in terms of the proportion of instances to which the rule applies. The third measure is **Lift**, i.e., the propensity to occur relative to average. Pairs of attributes are called itemsets (which in grocery terms could be products purchased together). Association rule analysis seeks itemsets above a specified support and confidence. The general approach is to identify itemsets with required coverage and turn each into a rule or set of rules with accuracy above another specified level. Association rules are hard to control, i.e., one of the unsupervised machine learning techniques – some itemsets will not produce any rules, others may produce many.

Association rules can provide information that can be useful in a number of different ways. In the field of marketing, for instance, they can provide:

- Identification of products to be placed together to attract customers for cross-selling;
- Targeting customers through marketing campaigns (coupons, mailings, e-mailings, etc.) seeking to get them to expand the number of products purchased;
- Drive recommendation engines in on-line marketing.

It is machine learning that has proven very useful to Amazon in recommending purchases based upon past purchases, as well as retail stores seeking to attract customers to purchase products based on current purchases. Thus, grocery stores locate things like bacon close to eggs and orange juice, or diapers with milk.

Outside of retailing, there are other uses for association rules. Classically, they were applied to retail transaction analysis, akin to market basket analysis. With the emergence of big data, the ability to apply association rules to streams of real-time data is highly useful, enabling a great deal of Web mining for many applications, including e-business retail sales. Association rule mining is one of the most widely used data mining techniques. This can be applied to target marketing, by customer profile, space allocation strategy within stores, but can also be extended to business applications such as international trade and stock market prediction. In science and engineering applications, remotely sensed imagery data has been analyzed to aid precision agriculture and resource discovery (to include oil). It has been used in manufacturing to analyze yield in semiconductor manufacturing. It has been used to improve the efficiency of packet routing over computer networks. In medicine it has been used for diagnosis of diseases. They also could be used in human resources management and other places where pairing behavior with results is of interest. In this chapter, we will focus on applications in healthcare, to follow the theme of the book.

4.5.1 The Apriori Algorithm

The apriori algorithm is credited to Agrawal, Imieliński, and Swami (1993) who applied it to market basket data to generate association rules. Association rules are usually applied to binary data, which fits the context where customers either purchase or don't purchase particular products. The apriori algorithm operates by systematically considering combinations of variables and ranking them on either support, confidence, or lift at the user's discretion.

The apriori algorithm operates by finding all rules satisfying minimum confidence and support specifications. First, the set of frequent 1-itemsets is identified by scanning the database to count each item. Next, 2-itemsets are identified, gaining some efficiency by using the fact that if a 1-itemset is not frequent, it can't be part of a frequent itemset of larger dimension. This continues to larger-dimensioned itemsets until they become null. The magnitude of effort required is indicated by the fact that each dimension of itemsets requires a full scan of the database. The algorithm is:

To identify the candidate itemset C_k of size k

1. Identify frequent items L_1

    For $k = 1$ generate all itemsets with support ≥ Support_{min}

    If itemsets null, STOP

        Increment k by 1

        For itemsets of size k identify all with support ≥ Support_{min}

    END

2. Return list of frequent itemsets
3. Identify rules in the form of antecedents and consequents from the frequent items
4. Check confidence of these rules

If confidence of a rule meets Confidence_{min} mark this rule is strong.

The output of the apriori algorithm can be used as the basis for recommending rules, considering factors such as correlation, or analysis from other techniques, from a training set of data. This information may be used in many ways, including in retail where if a rule is identified indicating that purchase of the antecedent occurred without that customer purchasing the consequent, then it might be attractive to suggest purchase of the consequent.

The apriori algorithm can generate many frequent itemsets. Association rules can be generated by only looking at frequent itemsets that are strong, in the sense that they meet or exceed both minimum support and minimum confidence levels. It must be noted that this does not necessarily mean such a rule is useful, that it means high correlation, nor that it has any proof of causality. However, a good feature is that you can let computers loose to identify them (an example of machine learning).

4.5.2 Association Rules with R

The R software allows setting support and confidence levels, as well as the minimum rule length. It has other options as well. We will set support and confidence (as well as lift, which is an option for sorting output) below. The data needs to be put into a form software can read. In Rattle, i.e., an R package for datamining with GUI requires data to be categorical rather than numerical. The rules generated will be positive cases (IF you buy diapers THEN you are likely to buy baby powder) and negative cases are ignored (IF you **didn't** buy diapers THEN you are likely to do whatever). If you wish to study the negative cases, you would need to convert the blank cases to No. Here we will demonstrate the positive case. Association rule mining seeks all rules satisfying specified minimum levels. Association rules require nominal data.

4.5.3 A Supply Chain Case and Dataset

For simple demonstration purposes, we will use a supply chain case and a relevant dataset. This dataset is posted on Kaggle.com (DataCoSupplyChainDataset.csv). It covers data for a supply chain firm over the period 2015 through 2018. The analysis revealed significant insights and actionable recommendations across various facets of the supply chain, from shipping delays and fraud detection to customer segmentation and market expansion. Key findings highlighted challenges in 'Same Day' shipping, potential for centralized warehouses, and opportunities in product profitability. It

TABLE 4.4

DataCoSupplyChainDataset Order Status Outcomes

Outcome	Number
Cancelled	3,692
Complete	59,491
Suspected Fraud	4,062
Pending	20,227
Processing	21,902
Pending Payment	39,832
On Hold	9,804
Payment Review	1,893

contains 180,519 observations off 117 stock keeping units with 65,752 unique orders in 50 product categories. There are 164 countries involved with 180,519 products ordered using one of four delivery modes. Association rules require categorical data. Of the 53 variables in the DataCoSupplyChainDataset.csv, 9 are categorical. We select the target variable to be Order Status. The other variables selected are:

• Type	4 categories
• Delivery Status	4 categories
• Market	5 categories
• Order Country	164 countries
• Order State	1,089 states
• Order Status	9 categories
• Product Name	118 products
• Ship Mode	4 categories

We did not select Order City as there were too many. Order Status outcomes are given in Table 4.4.

4.5.4 R Association Rules

The first step is to install the R rules package, and access it with the library statement:

```
#Association Rules
install.packages("arules", repos = "http://cran.us.r-proj
ect.org")

library(arules)
```

The syntax to access the data file is given here:

```
setwd("C:/Users/oaraz2/Desktop")
SCData = as.data.frame(read.csv("DataCoSupplyChainAssoci
ation.csv"))
sum(is.na(SCData))
```

The following warning was obtained as the data was categorical. The data was discretized:

```
## [1] 0

SC_trans=as(SCData, "transactions")

## Warning: Column(s) 1, 2, 3, 4, 5, 6, 7, 8 not logical
or factor. Applying
## default discretization (see '? discretizeDF').
```

Column names are accessed from the data file. The apriori algorithm is accessed with the rules line, setting support limit of 0.1 and confidence at 0.01 for purposes of demonstration. These two parameters (support and confidence) are the controls to adjust the number of rules obtained. It is almost always a matter of experimentation, as some settings will return no rules, others might return thousands or more.

```
#inspect(SC_trans)

col_names <- names(SCData)
SCData[,col_names] <- lapply(SCData[,col_names], factor)
rules=apriori(SCData, parameter=list(supp=0.1, conf=0.01,
target="rules",minlen=2))
```

The output for this set of support and confidence returns 70 rules. The initial output shows parameter settings. The inspect(sort(rules, by "lift")) can be adjusted to sort rules obtained by lift, support (joint), confidence, or coverage (support of antecedent).

```
## Apriori
##
## Parameter specification:
##   confidence minval smax arem  aval originalSupport
maxtime support minlen
##         0.01    0.1    1 none FALSE            TRUE
5     0.1      2
##  maxlen target  ext
##      10  rules TRUE
##
## Algorithmic control:
##  filter tree heap memopt  load sort verbose
##     0.1 TRUE TRUE  FALSE  TRUE    2    TRUE
##
## Absolute minimum support count: 18051
##
## set item appearances ...[0 item(s)] done [0.00s].
## set transactions ...[1397 item(s), 180519 transaction(s)]
done [0.24s].
## sorting and recoding items ... [24 item(s)] done [0.01s].
## creating transaction tree ... done [0.06s].
## checking subsets of size 1 2 3 done [0.00s].
## writing ... [70 rule(s)] done [0.00s].
## creating S4 object  ... done [0.01s].

inspect(sort(rules, by="lift"))
```

The first four of the 70 rules obtained were:

	Lhs		rhs	support	confidence	coverage	lift	count
[1]	{OrdStatus= CLOSED}	=>	{Type=CASH}	0.109	1.000	0.109	9.203	19616
[2]	{Type=CASH}	=>	{OrdStatus= CLOSED}	0.109	1.000	0.109	9.203	19616
[3]	{OrdCountry= Estados Unidos	=>	{Market=USCA}	0.138	1.000	0.138	6.997	24840
[4]	{Market=USCA}	=>	{OrdCountry= Estados Unidos}	0.138	0.963	0.143	6.997	24840

The first rule's antecedent is that if order status is CLOSED, then payment was in cash. This occurred in 10.9 percent of the training cases, and if order status was closed, all payments were cash (confidence was 1.0). Coverage was

TABLE 4.5

Number of Association Rules by Support/Confidence Levels

Support	Confidence	# Rules
0.01	0.1	70
0.009	0.09	88
0.008	0.08	101
0.007	0.07	180
0.006	0.06	270
0.05	0.05	417
0.01	0.01	6,328

10.9 percent which follows as all of the antecedent cases paid cash. This had a very high lift of 9.2. This combination occurred 19,616 cases in the training set. The second rule is the corresponding complement of the first rule. If payment was cash, order status was closed, based on the same 19,616 case.

The third rule was that if the order came from the United States, market was USCA. This occurred 24,840 times, with support of 0.138 and confidence of 1.0. Coverages match support here. The fourth rule is the complement of the third rule with the same count of 24,840, same support, but less confidence as there were cases where the market was USCA and the order did not come from the United States.

The user can control the number of rules by adjusting the support and confidence parameters. In the above output, these were both set to 0.1. Table 4.5 shows the impact of varying these parameters.

4.6 Conclusions

Supply chains involve large risks inherently due to the need for coordination across supplies, logistics, production, and retail. This chapter began with reviewing generic supply chain risks. Risk management frameworks were then discussed. This was followed by brief presentation of some tools that could be used to better manage supply chain risk. Risk matrices provide a simple means to implement risk management policies. Multiple criteria frameworks give a means to aid in selection of options. Balanced scorecards provide a means to monitor supply chain risk management performance. Mitigation strategies were reviewed, providing options for dealing with supply chain risk management problems. Finally, information technology can be applied. While the costs of application remain to be thoroughly examined, blockchain technology is being applied to enhance supply chain risk management operations.

References

Agrawal R, Imielinski T, Swami A (1993) Mining association rules between sets of items in large databases. In: Buneman P, Jajodia S (eds) *Proceedings of the 1993 ACM SIGMOD international conference on management of data,* 201–216. New York: Association of Computer Machinery.

Baryannis G, Validi S, Dani S, Antoniou G (2019) Supply chain risk management and artificial intelligence: State of the art and future research directions, *International Journal of Production Research* 57(7): 2179–2202.

Buchholz P, Schumacher A, Al Barazi S (2022) Big data analyses for real-time tracking of risks in the mineral raw material markets: Implications for improved supply chain risk management, *Mineral Economics* 35: 701–744.

Chowdhury S, Rodriguez-Espindola O, Dey P, Budhwar P (2022) Blockchain technology adoption for managing risks in operations and supply chain management: Evidence from the UK, *Annals of Operations Research* 327: 539–574.

DuHadway S, Carnovale S, Hazen B (2019) Understanding risk management for intentional supply chain disruptions: Risk detection, risk mitigation, and risk recovery, *Annals of Operations Research* 283: 179–198.

Ellis SC, Rao S, Raju D, Goldsby TJ (2018) RFID tag performance: Linking the laboratory to the field through unsupervised learning, *Production and Operations Management* 27(10): 1834–1848.

González-Zapatero C, González-Benito J, Lannelongue G, Miguel Ferreira L (2021) Using fit perspectives to explain supply chain risk management efficacy, *International Journal of Production Research* 59(17): 5272–5283.

Jain V, Wadhwa S, Deshmukh SG (2007) Supplier selection using fuzzy association rules mining approach, *International Journal of Production Research* 45(6): 1323–1353.

Joseph Jerome JJ, Sonwaney V, Bryde D, Graham G (2023) Achieving competitive advantage through technology-driven proactive supply chain risk management: An empirical study, *Annals of Operations Research* 332: 149–190.

Kaplan RS, Norton DP (1996) *The Balanced Scorecard: Translating Strategy into Action.* Boston, MA: Harvard Business School Press.

Lee CKH, Ho GTD, Choy KL, Pang GKH (2014) A RFID-based recursive process mining system for quality assurance in the garment industry, *International Journal of Production Research* 52(14): 4216–4238.

Mizgier KJ, Kocsis O, Wagner SM (2018) Zurich Insurance uses data analytics to leverage the BI insurance proposition, *Interfaces* 48(2): 94–107.

Olson DL, Chae B (2023) Incorporating an unsupervised text mining approach into studying logistics risk management: Insights from corporate annual reports and topic modeling, *Information* 14: 395–413.

Olson DL, Wu D (2023) *Enterprise Risk Management Models: Focus on Sustainability,* fourth edition. Berlin: Springer.

Shareef MA, Dwivedi YK, Kumar V, Hughes DL, Raman R (2020) Sustainable supply chain for disaster management: Structural dynamics and disruptive risks, *Annals of Operations Research* 319: 1451–1475.

Ting SL, Tse YK, Ho GTS, Chung SH, Pang G (2014) Mining logistics data to assure the quality in a sustainable food supply chain: A case in the red wine industry, *International Journal of Production Economics* 152: 200–209.

Tsang YP, Wu CH, Lin K-Y, Tse YJ, Ho GTS, Lee CKM (2022) Unlocking the power of big data analytics in new product design framework in the furniture industry, *Journal of Manufacturing Systems* 62: 777–791.

U.S. Department of Commerce, National Institute of Standards and Technology (2018) *Risk Management Framework for Information Systems and Organizations: A System Life Cycle Approach for Security and Privacy*. https://doi.org/10.6028/nist.sp.800-37r2.

Wang J (2024) Strategic analysis of biosafety in agricultural product supply chain management within e-commerce platforms, *Journal of Commercial Biotechnology* 29(1): 124–137.

Wang J, Zhao L, Huchzermeier A (2020) Operations-finance interface in risk management: Research evolution and opportunities, *Production and Operations Management* 30(2): 355–389.

Wicaksana A, Ho W, Talluri S, Dolgui A (2022) A decade of progress in supply chain risk management: Risk typology, emerging topics, and research collaborators, *International Journal of Production Research* 60(24): 7155–7177.

5

Simulating Supply Chain Risks

This chapter presents Monte Carlo simulation models related to supply chain management. Simulation is widely used when probability distributions as well as some degree of complexity are present. Monte Carlo simulation involves building a model consisting of events with probabilities distributions, and instead of seeking to develop a closed-form analytic model to optimize, draw random events following the prescribed distributions to measure system outputs. It generates statistics describing expected system performance. It has proven highly useful in supply chain management research and operations management, e.g., both for waiting line systems, and supply chain and/or financial risks. Monte Carlo models are used to predict possible outcomes of uncertain events. Recent interesting supply chain simulation studies have included technical-economic analysis of bioethanol production in Iran (Hasanly et al., 2018), prediction of El Niño oscillation (Wang et al., 2019), and selection of best ports in the US Gulf Coast (De Icaza et al., 2019). There are also some Monte Carlo models published which have included analysis of CO_2 emission (Hong et al., 2019), environmental reuse in a supply chain (Liao et al., 2020), inventory stocking policy evaluation (Ghosh et al., 2022), and supply chain network risk management (Qazi et al., 2023). Freitas Silva et al. (2021) presented a Crystal Ball model for risk assessment of gas supply in Brazil. Ghadge et al. (2022) gave a system dynamics simulation model of a supply chain network's COVID-19 disruptions. Simulation is widely used because of its flexibility. The downside is that it is a statistical analysis, without concrete conclusions for the most part.

This chapter will present supply chain decision problems with simulation models. The first model relates to the vendor selection problem presented in Chapter 4. The second examines financial net present value of a supply chain information systems project. The third involves a traditional inventory ROP/quantity ordered model. The fourth presents a simple system dynamics simulation of a supply chain inventory system.

 DOI: 10.1201/9781003562399-5

5.1 Vendor Selection Monte Carlo Model

There are six vendors being considered, each of which has given a quoted unit price for a part that is needed on a large volume. Risks are given as probabilities of not failing each particular risk category. Simulation is run to generate 1,000 samples of actual realized price after exchange rate variance, to include having to rely upon an expensive ($10 per unit) price in case of outsourcing vendor failure. There are two Chinese vendors who use the same currency, so you only have to simulate one exchange rate shared by both. There are two US vendors whose currency is already in dollars. Data for the Monte Carlo simulation is given in Table 5.1.

In supply chain outsourcing decisions, a number of factors can involve uncertainty, and simulation can be useful in gaining better understanding of systems. We begin by looking at expected distributions of prices for the component to be outsourced from each location, which empirically is at least close to normal, and we have data on average inflation and standard deviation. The four risk probabilities are shown, with overall risk being the product of the four. If a vendor fails for any of the four reasons, an alternative price of $10 per unit would be incurred.

Monte Carlo simulation is applied to obtain expected performance. Simulation modeling is performed using Crystal Ball software. Other software such as @Risk works very much the same. Our purpose is not to cover Crystal Ball, so we will focus on the output obtained. The Crystal Ball input in Excel format is given in Table 5.2.

For each vendor, the quoted price is inflated using a normal distribution with mean inflation/month as given in Table 5.1 along with its standard deviation, along with a $10 price if any of the four risk factors fail.

Output involves identifying contract fulfillment which is equal to the product of the four risks, and realized cost, which is the contract fulfillment probability times the realized cost + $10 (the alternative cost in case of contract failure) times 1–contract fulfillment probability. The results from

TABLE 5.1

Vendor Simulation Data

Location	Quoted	StdDev	Product Risk	Firm Risk	Country Risk	Disaster
Alabama	$5.50	0	0.85	0.93	0.99	0.96
Oregon	$5.40	0	0.90	0.90	0.99	0.98
Mexico	$3.40	2.21	0.75	0.92	0.85	0.93
Germany	$7.50	0.05	0.99	0.99	0.97	0.99
China1	$3.80	0.32	0.90	0.85	0.90	0.92
China2	$4.10	0.32	0.95	0.93	0.90	0.94

TABLE 5.2
Crystal Ball Vendor Model

A	B	C	D	E	F	G	H
1 **Loc**	**Bid**	**StdD**	**Realized**	**Product Risk**	**Firm Risk**	**Country Risk**	**Disaster**
2 Alabama	$5.50	0	Normal(B2,C2)	YesNo(0.85)	YesNo(0.93)	YesNo(0.99)	YesNo(0.96)
3 Oregon	$5.40	0	Normal(B3,C3)	YesNo(0.90)	YesNo(0.90)	G2	YesNo(0.98)
4 Mexico	$3.40	2.21	Normal(B4,C4)	YesNo(0.75)	YesNo(0.92)	YesNo(0.85)	YesNo(0.93)
5 Germany	$7.50	0.05	Normal(B5,C5)	YesNo(0.99)	YesNo(0.99)	YesNo(0.97)	YesNo(0.99)
6 China1	$3.80	0.32	Normal(B6,C6)	YesNo(0.90)	YesNo(0.85)	YesNo(0.90)	YesNo(0.92)
7 China2	$4.10	0.32	Normal(B7,C7)	YesNo(0.95)	YesNo(0.93)	G6	YesNo(0.94)

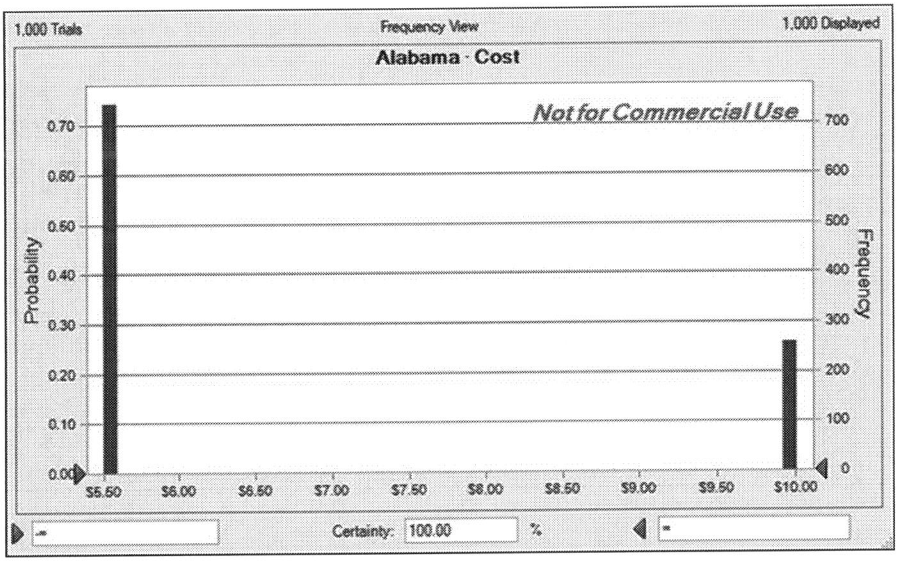

FIGURE 5.1
Alabama Cost.

the Monte Carlo simulation model based on 1,000 replications are given in Figures 5.1 through 5.6.

Figures 5.1 through 5.6 show the simulated costs. The US vendors had no currency variance, so the price was either the bid quote or the default $10 (for those cases where the contract was not fulfilled due to various risks). Mexico had a highly variable (non-normal) distributio n but the other three a normal kind of distribution for those cases where the contract did not fail, along with $10 costs in case of failure. Relative costs are displayed in Table 5.3.

Based on mean unit costs, China2 vendor is the best choice at $5.52. The histogram in Figure 5.6 shows a probability of 0.24 of contract failure, but at $10 for an alternative supply, that still is economical compared to the other

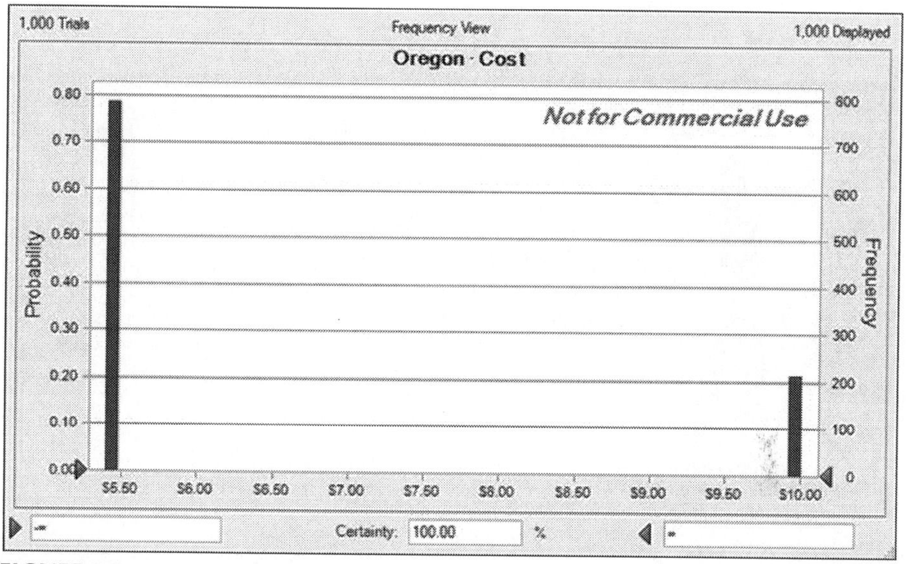

FIGURE 5.2
Oregon Cost.

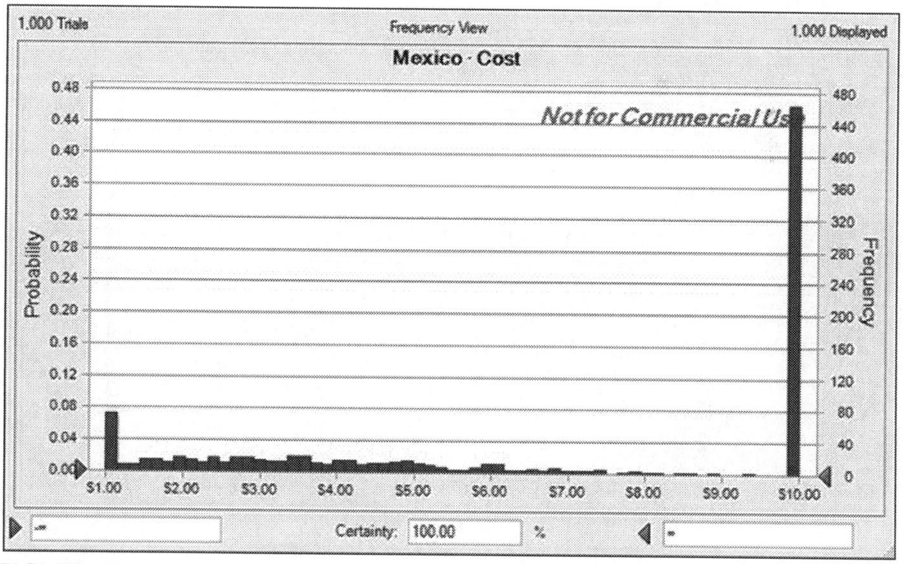

FIGURE 5.3
Mexico Cost.

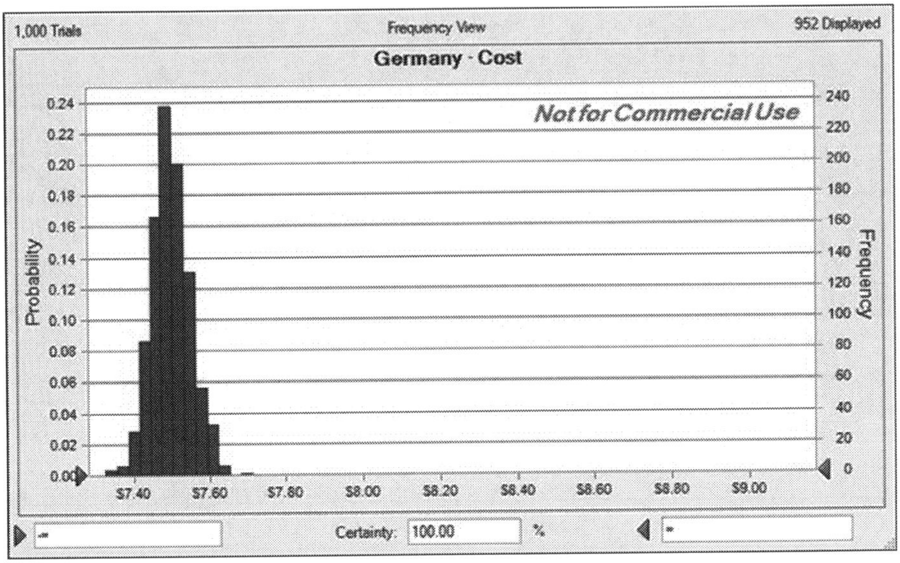

FIGURE 5.4
Germany Cost.

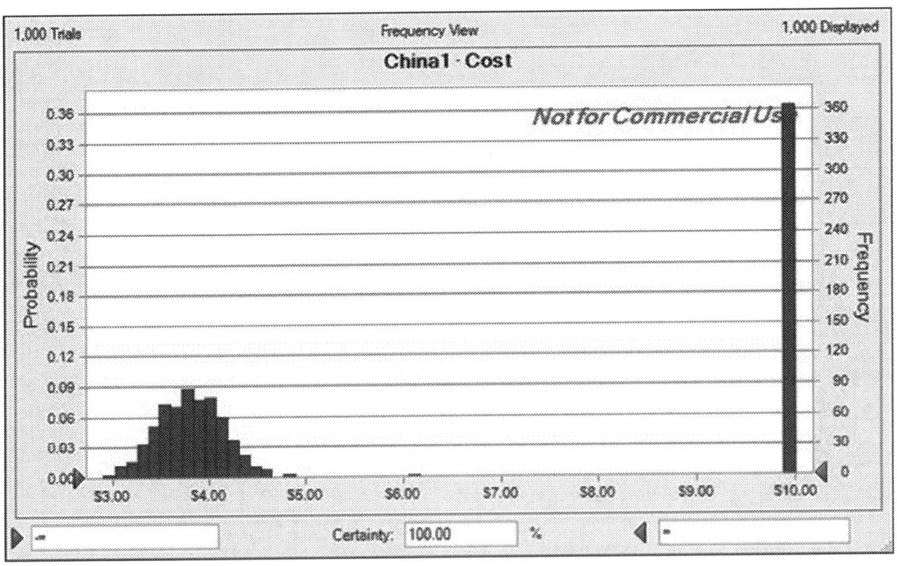

FIGURE 5.5
China1 Cost.

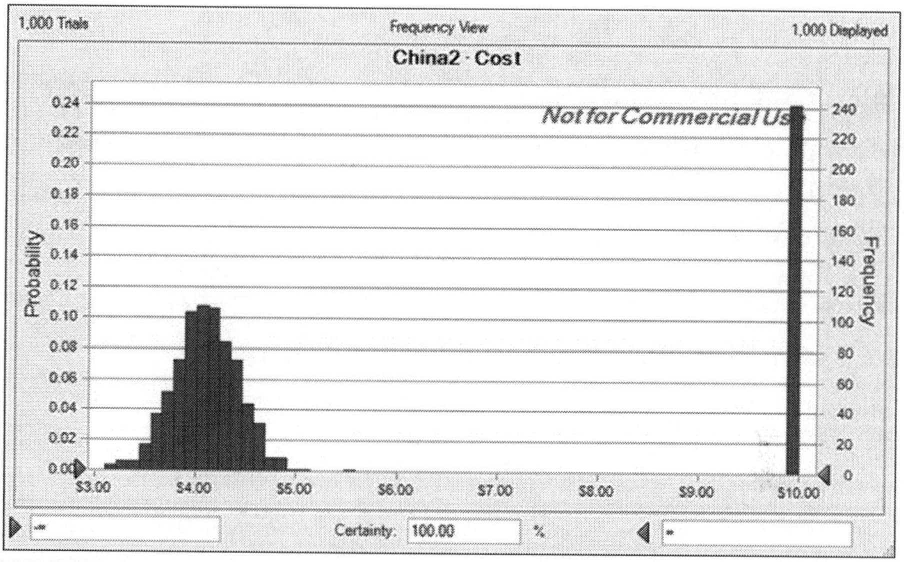

FIGURE 5.6
China2 Cost.

TABLE 5.3
Monte Carlo Vendor Cost Results

Statistic	Alabama	Oregon	Mexico	Germany	China1	China2
Trials	1,000	1,000	1,000	1,000	1,000	1,000
Base Case	$6.62	$6.39	$6.40	$7.65	$6.07	$5.59
Mean	**$6.66**	**$6.39**	**$6.50**	**$7.62**	**$6.07**	**$5.52**
Median	$5.50	$5.40	$6.62	$7.50	$4.09	$4.23
Mode	$5.50	$5.40	$10.00	$10.00	$10.00	$10.00
Standard Dev	$1.97	$1.89	$3.53	$0.54	$2.99	$2.54
Variance	$3.88	$3.56	$12.46	$0.29	$8.93	$6.46
Skewness	1.1	1.39	(0.23)	4.17	0.54	1.17
Kurtosis	2.22	2.94	1.36	18.54	1.32	2.43
Coeff. of Variation	0.2956	0.2956	0.5429	0.0705	0.4926	0.4603
Minimum	$5.50	$5.40	$1.00	$7.33	$2.81	$3.11
Maximum	$10.00	$10.00	$10.00	$10.00	$10.00	$10.00
Mean Std. Error	$0.06	$0.06	$0.11	$0.02	$0.09	$0.08

five vendors. Figures 5.1 and 5.2 show Alabama and Oregon plants to have no variance in quote, but some probabilities of failure to fulfill the contract. Mexico has a good bid price, but too high a probability of failure. Germany has few failures but is too expensive. China1 vendor has more failures than China2.

China1 in this case has the lowest estimated price, but it has a wide expected distribution of exchange rate fluctuation. These distributions will affect the actual realized price for the outsourced component. The China1 vendor is also rated as having relatively high probabilities of failure in product compliance with contractual standards, in vendor financial survival, and in political stability of host country.

5.2 Financial Simulation of Net Present Value

Consider a supply chain software installation with the following conditions (see Table 5.4).

Cost of capital 20 percent per year	
Invest at year 0:	3 million consultant
Invest at year 1:	2 million consultant
	7 million vendor
	3 million hardware
Invest at year 2:	1 million consultant
	1 million hardware
Invest at year 3:	1 million consultant

Project investments are constants. All of the operating and training expenses as well as benefits are distributions. We calculate NPV for this supply chain software project based on 20 percent marginal value of capital. Figure 5.7 gives resulting output.

TABLE 5.4

Financial Simulation NPV Data

End of:	Project Investment	Operating Expense	Training Expense	Benefit
Year 0	3 million constant			
Year 1	12 million constant	1 million Normal(1,0.2)	0.1 million Normal(.1,.01)	
Year 2	2 million constant	2 million Normal(2,0.5)	0.5 million Normal(0.5,0.06)	1 million Exponential(1)
Year 3	1 million constant	2.5 million Normal(2.5,0.7)	1 Million Normal(1,0.08)	5 million Exponential(0.2)
Year 4		3 million Normal(3,0.9)	0.5 million Normal(0.5,0.07)	8 million Exponential(0.125)
Year 5		3.5 million Normal(3.5,1)	0.2 million Normal(0.2,0.03)	20 million Exponential(0.05)
Year 6		4 million Normal(4,1.2)	0.3 million Normal(0.3,0.03)	16 million Exponential(0.0625)

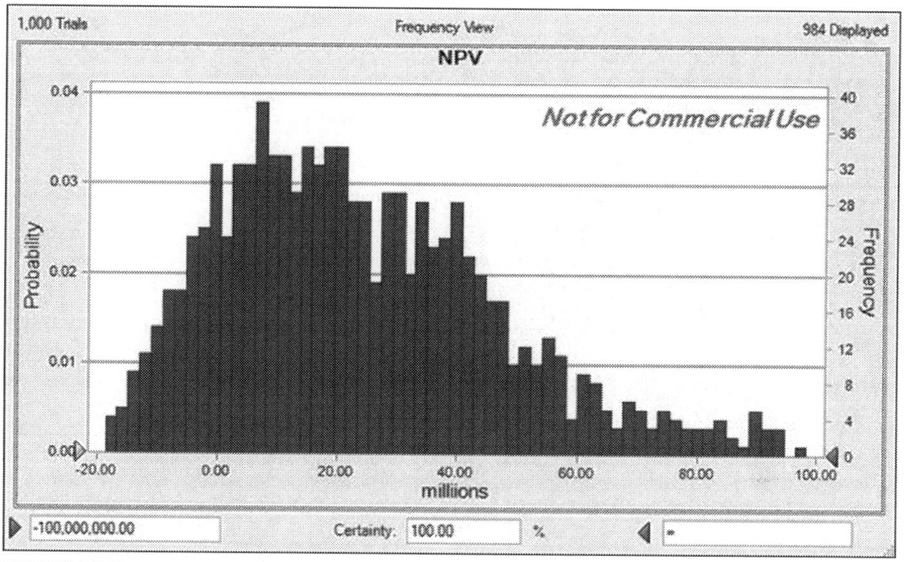

FIGURE 5.7
Forecast: NPV.

TABLE 5.5
NPV Statistics

Statistic	Forecast Values
Trials	1,000
Mean	25.19
Median	20.64
Standard Deviation	26.10
Variance	681.24
Skewness	1.26
Kurtosis	6.10
Coeff. of Variation	1.04
Minimum	–18.61
Maximum	192.03
Mean Std. Error	0.83

The statistics behind this result are given in Table 5.5.

The mean result was a net present value over 6 years of $25.19 million, indicating an attractive investment. The odds of NPV benefits minus costs being at least $1 million are displayed in Figure 5.8. The Monte Carlo simulation shows that while the NPV is well over $25 million, there is a 17 percent chance that the NPV would turn out negative. Figure 5.9 shows similar results for results breaking even.

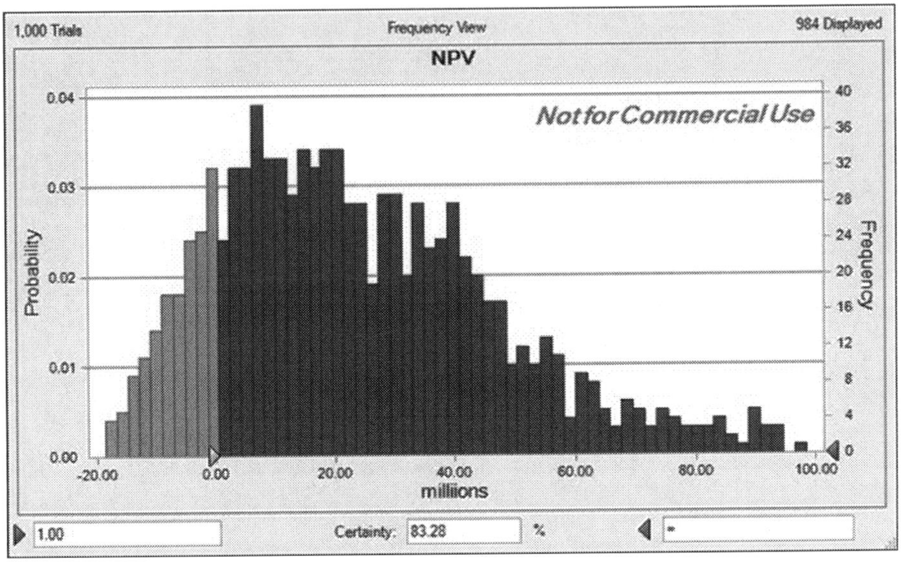

FIGURE 5.8
Histogram Displaying Results over $1 Million.

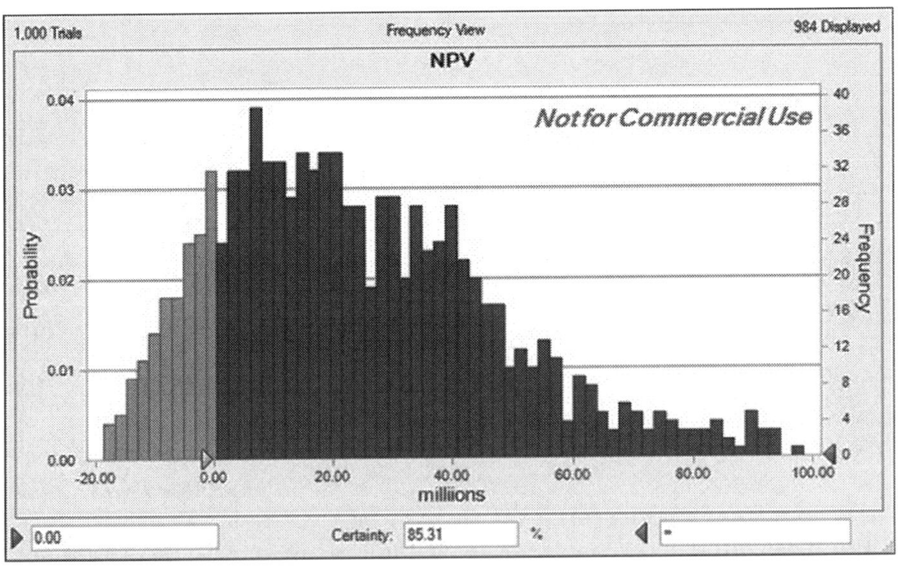

FIGURE 5.9
Monte Carlo NPV Results Breaking Even.

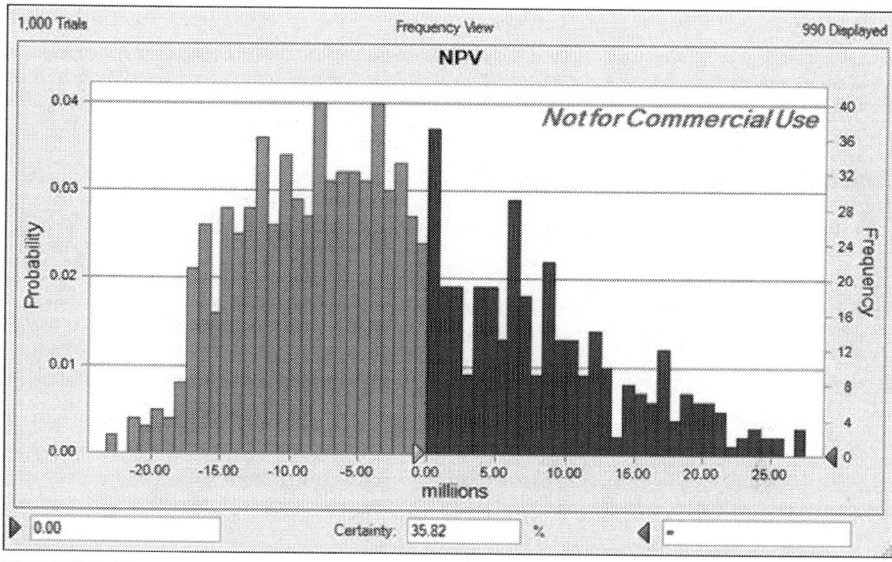

FIGURE 5.10
NPV with Reduced Benefits.

TABLE 5.6
Conservative NPV Statistics

Forecast	NPV
Base Case	–1.26
Mean	–2.31
Median	–3.86
Standard Deviation	10.67
Variance	113.78
Skewness	0.7986
Kurtosis	3.65
Coeff. of Variation	–4.61
Minimum	–23.33
Maximum	50.37
Mean Std. Error	0.34

Figure 5.9 shows a probability of 0.85 of having an NPV over 6 years of at least breaking even. A more conservative estimate of benefits would be entirely reasonable. Assume that estimated benefits were reduced to 4 million in year 3, 5 million in year 4, 8 million in year 5, and 5 million in year 6. The model can be easily rerun with different benefits in years 5 and 6 yielding Figure 5.10.

Figure 5.10 shows a probability of 0.36 of a positive NPV. Table 5.6 gives results for this run.

The point is that results are highly sensitive to assumed values. Nonetheless, Monte Carlo simulation offers a valuable means to model expected results.

5.3 Monte Carlo Simulation of an Inventory Model

Consider an inventory situation where demand is exponentially distributed with a mean demand of 100 per day (an integer variable). Starting inventory is 100, sales price is $500, purchase price is $100, and holding cost is $10/item in inventory at the end of each day, Ordering cost is $50 per order, and what is ordered at the end of each day is received before work the next morning. We model profit as well as unfilled orders with the intent of comparing policies for quantity ordered and reorder point.
We begin by calculating EOQ to get:

$$EOQ = \sqrt{\frac{2*100*360*50}{10}} = 600$$

We model 30 days and simulate that 1,000 times.

Start	Demand	Sales	End	Order	HC	OC	PC	Rev	Short
1 100	expon (100)	min(start, demand)	start-sales	IF(Q,0)	10*end	50*order	100* order	500* sales	demand-sales
2 prior end	expon (100)	min(start, demand)	start-sales	IF(Q,0)	10*end	50*order	100* order	500* sales	demand-sales

This model yields the results given in Table 5.7.

Here the optimal outcome of ROP=260 and Q=220 yielded a net profit of $1,053,381. Note that this type of model notoriously has a very flat surface

TABLE 5.7
Inventory Model Results

	Q(200)	Q(220)	Q(240)	Q(260)	Q(280)	Q(300_
ROP (240.00)	1,048,195	1,049,924	1,052,393	1,051,393	1,052,698	1,051,942
ROP (250.00)	1,049,911	1,052,413	1,052,260	1,051,771	1,052,679	1,052,511
ROP (260.00)	1,050,185	**1,053,381**	1,051,639	1,052,104	1,051,847	1,051,480
ROP (270.00)	1,050,463	1,052,976	1,051,690	1,051,618	1,052,674	1,051,363
ROP (280.00)	1,051,308	1,052,252	1,051,462	1,052,013	1,052,744	1,051,551

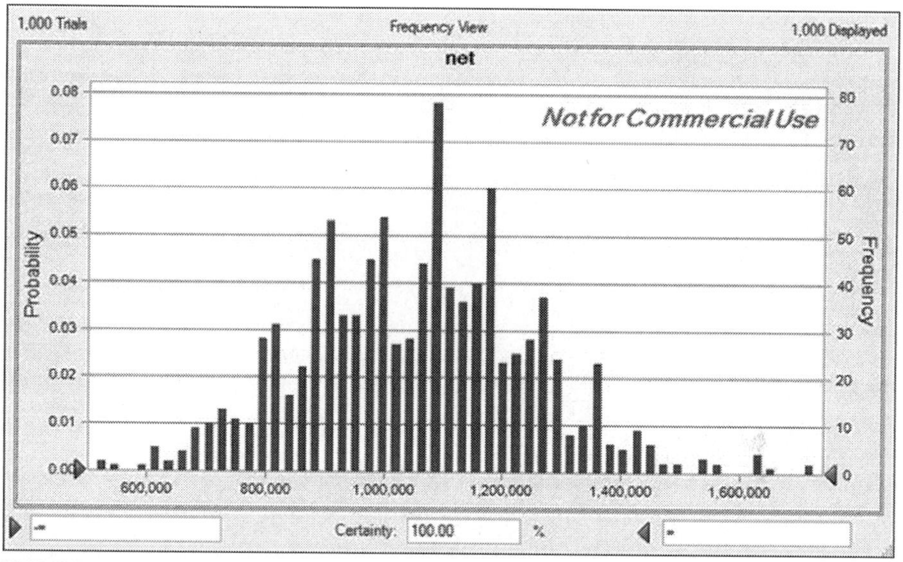

FIGURE 5.11
Inventory Net Profits.

TABLE 5.8
Inventory Model Statistics

Forecast: Rev.	Revenue	PC	OC	HC	Net
Base Case	1,500,000.00	330,000.00	750.00	73,200.00	1,104,050.00
Mean	1,424,247.00	310,816.00	706.40	75,850.32	1,053,380.88
Median	1,436,500.00	308,000.00	700.00	76,000.00	1,063,470.00
Mode	1,590,000.00	330,000.00	750.00	73,670.00	866,420.00
Standard Deviation	231,595.55	46,482.41	105.64	7,029.77	189,564.17
Skewness	0.17	0.1499	0.1499	−0.2944	0.1781
Kurtosis	3.09	3.07	3.07	3.24	3.09
Coeff. of Variation	0.1626	0.1495	0.1495	0.0927	0.18
Minimum	758,000.00	176,000.00	400.00	49,650.00	513,420.00
Maximum	2,270,500.00	484,000.00	1,100.00	94,660.00	1,730,350.00
Mean Std. Error	7,323.69	1,469.90	3.34	222.30	5,994.55

(in other words, you can be off and still come close to the same best result). Figure 5.11 displays the histogram if simulated profits.

Table 5.8 gives the monthly revenues, purchase costs, order costs, holding costs, and net for the simulation.

Figure 5.12 displays the quantity short per month for this model. Statistics for shortfall are given in Table 5.9.

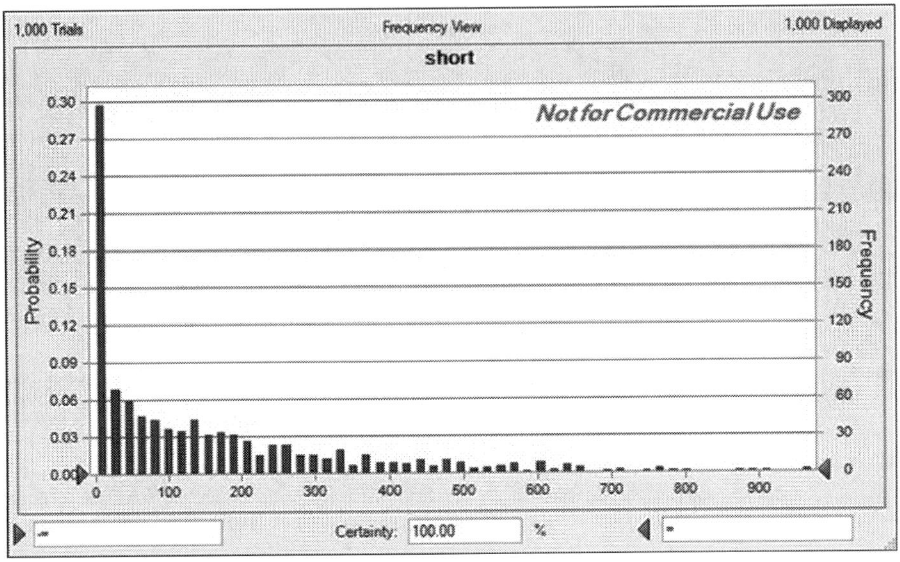

FIGURE 5.12
Inventory Model Shortfall Histogram.

TABLE 5.9
Inventory Model Shortfall Statistics

Forecast: short	
Base Case	0
Mean	146.25
Median	81.5
Mode	0
Standard Deviation	175.52
Skewness	1.62
Kurtosis	5.63
Coeff. of Variation	1.2
Minimum	0
Maximum	967
Mean Std. Error	5.55

Running the model using the EOQ of 600 and ROP equal to the expected demand during lead time of one day (100) yields the histogram shown in Figure 5.13, while statistics are given in Table 5.10.

The net gain for the EOQ model is much less than that identified with the Monte Carlo decision table. Shortages are displayed in Figure 5.14.

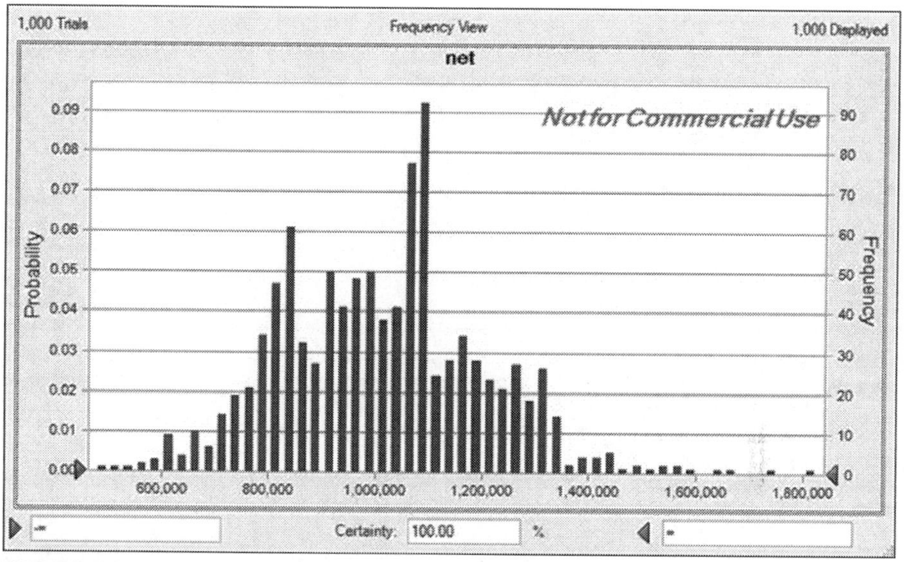

FIGURE 5.13
EOQ Simulation Net.

TABLE 5.10
EOQ Simulation Statistics

	Revenue	PC	OC	HC	Net	Short
Base Case	1,500,000	300,000	250	75,000	1,124,750	0
Mean	1,396,384	310,320	259	95,639	1,011,369	202
Median	1,387,250	300,000	250	95,295	1,013,970	156
Mode	1,550,000	300,000	250	97,670	818,920	0
Standard Deviation	234,787.11	51,374.48	42.81	9,251.43	189,643.51	189.44
Skewness	0.3142	0.266	0.266	0.2418	0.2971	1.25
Kurtosis	3.35	3.17	3.17	3	3.34	4.38
Coeff. of Variation	0.1681	0.1656	0.1656	0.0967	0.1875	0.9379
Minimum	724500	180000	150	68810	475060	0
Maximum	2396500	480000	400	125460	1826830	1037
Mean Std. Error	7,424.62	1,624.60	1.35	292.56	5,997.05	5.99

5.4 System Dynamics Inventory Modeling

Another form of inventory simulation is related to the beer game (Sterman, 2010), a popular role-playing exercise often used in operations management courses of MBA programs. Sterman used system dynamics modeling, supported by special software, but doable with spreadsheet modeling.

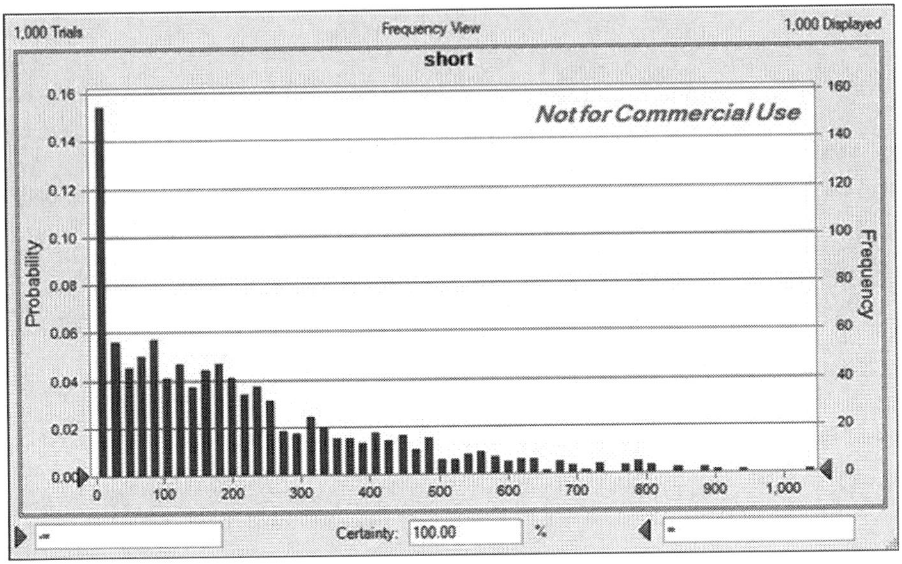

FIGURE 5.14
EOQ Shortages.

TABLE 5.11

Bottled Drink Inventory Costs

Level	Purchase	Order	Hold	Revenue
Factory	5	200	0.01	
Distributor	8	100	0.02	
Retailer	10	50	0.04	50

Ghadge et al. (2022) published an inventory model using system dynamics equations for a supply chain, studying the ripple effect of disruptions.

Consider a three-level system providing bottled drinks to convenience stores. We simplify to a system with one factory feeding two distribution warehouses that feed two retail stores each. Table 5.11 gives the inventory cost rates.

Both ROP and Q are set to 400 total for each level (so the two distributors have ROP of 200 each, the four retailers ROP of 100 each). Order rates are per unit. There is 1 factory, 2 distributors, and 4 retailers, so order costs are $200 per level in each time period. Delivery is done in every time period so ordering is a fixed cost of $600 per time period. Purchase costs are the start quantity for the first period, the order of the prior period for all subsequent periods. Holding cost is applied on ending inventory. Net is calculated using the retailer revenue. Table 5.12 gives spreadsheet equations for the first two periods.

TABLE 5.12

System Dynamics Spreadsheet Equations

	1	2
FactoryBegin	400	400
FactorySales	=I40+I41	=J40+J41
FactoryProd	=IF((I2–I3)<R7,R8,0)	=IF((J2–J3)<R7,R8,0)
WH1Begin	200	200
WH1Sales	=I35+I36	=MIN(J5,(I11+I15+I19+I23))
WH1Order	=IF((I5–I6)<R4,R5,0)	=IF((J5–J6)<R4,R5,0)
WH1End	=I5–I6	=J5–J6+I7
Ret1Begin	100	=I12+I31
Ret1Sales	=MIN(I9,I25)	=MIN(J9,J25)
Ret1Order	=IF((I9–I10)<R1,R2,0)	=IF((J9-J10)<R1,R2,0)
Ret1End	=I9–I10	=J9–J10+I11
Demand1	=INT(Exponential(100))	=INT(Exponential(100))

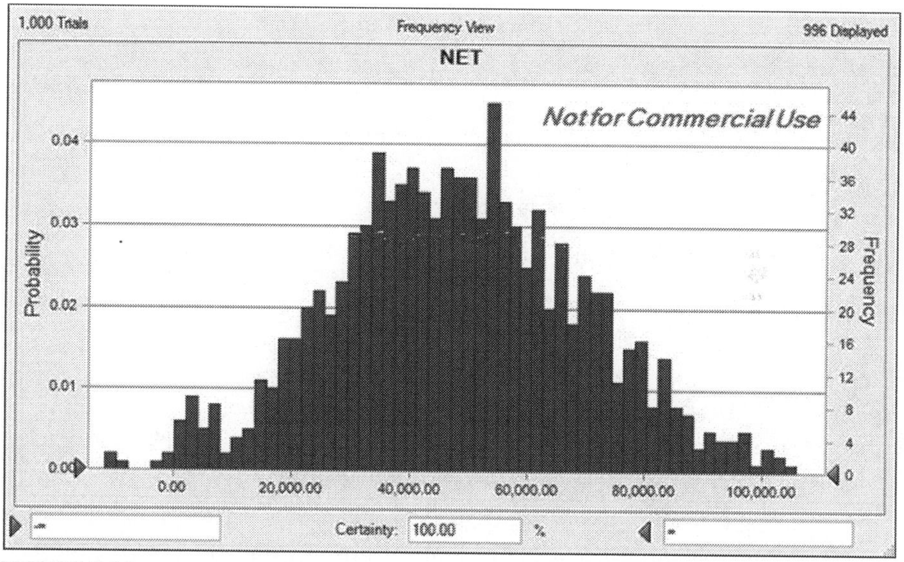

FIGURE 5.15
System Net Profit for Base Run.

The base run yielded net profits for the system shown in Figure 5.15.

The mean profit over 1,000 simulation runs was $48,062.97 over 13 time periods, with a minimum of $30,697 and a maximum of $115,401. The total of retail shortages is shown in Figure 5.16.

The mean shortage over 13 time periods was 902, with a minimum of 0 and a maximum of 2,624. The model was used to experiment with increasing Q per retailer per time period, yielding the net shown in Table 5.13.

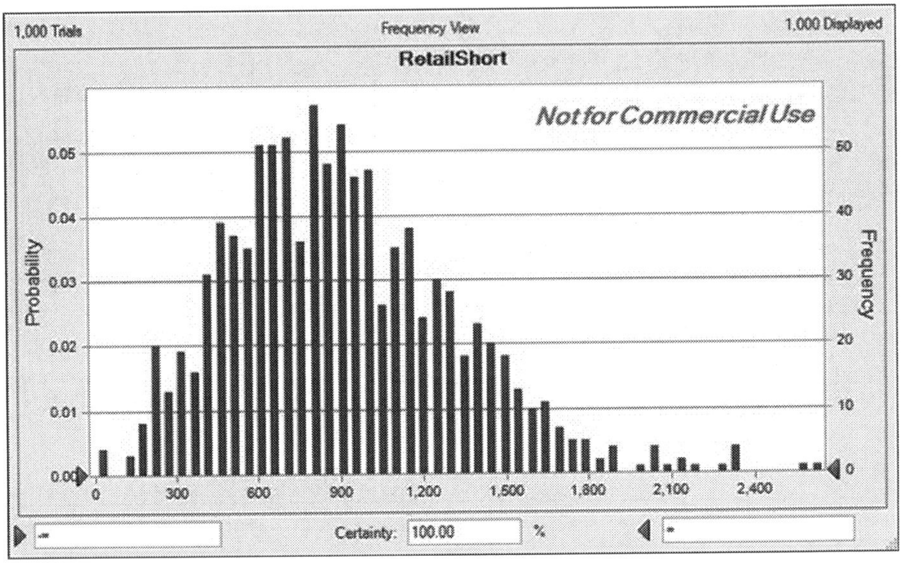

FIGURE 5.16
Total Retail Shortages.

TABLE 5.13
Results for Varying Retailer Q

Q per Retailer	Mean Net	Mean Short
100	48,062	902
110	49,371	855
120	50,382	816
130	51,591	775
150	53,654	703
200	56,167	586
300	52,040	496

While we didn't refine the study to find the optimal around Q of 200, Table 5.13 shows the trend. Retail shortages for Q of 200 are displayed in Figure 5.17.

Spreadsheet simulation models provide useful tools to experiment with input parameters.

5.5 Conclusions

Simulation is the most flexible management science modeling technique. It allows making literally any assumption you want, although the trade-off is

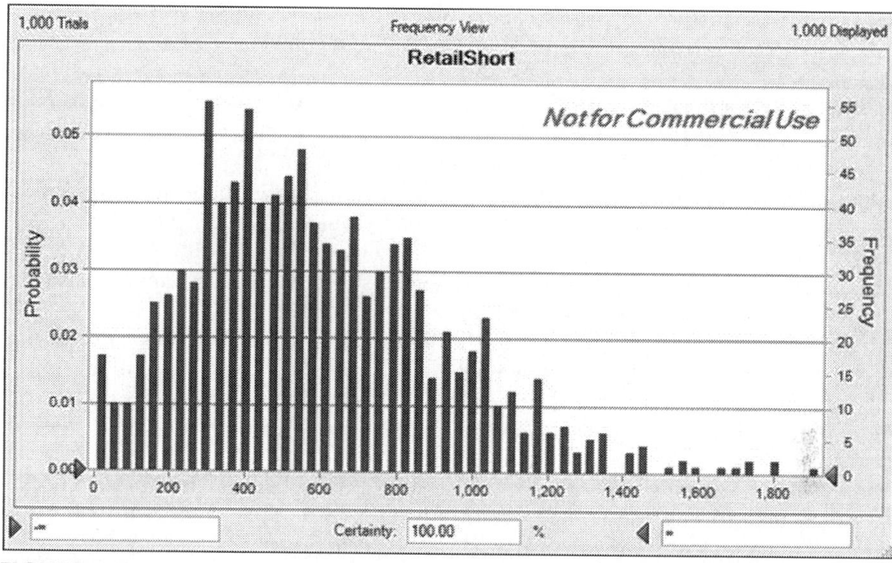

FIGURE 5.17
Retail Shortages with Q of 200.

that you have to work very hard to interpret results in a meaningful way relative to your decision.

Because of the variability inherent in risk analysis, simulation is an obviously valuable tool for risk analysis. There are two basic simulation applications in business. Waiting line models involve queuing systems, and waiting line model software is very appropriate for that type of modeling. The other type is supportable by spreadsheet tools such as Crystal Ball, demonstrated in this chapter. Spreadsheet simulation is highly appropriate for inventory modeling as in push/pull models. Spreadsheet models also are very useful for system dynamic simulations.

References

De Icaza RR, Parnell GS, Pohl EA (2019) Gulf coast port selection using multiple-objective decision analysis, *Decision Analysis* 16(2): 87–104.

Freitas Silva LM, Rodrigues de Oliveira AC, Alexandre Leite MS, Marins FAS (2021) Risk assessment model using conditional probability and simulation: Case study in a piped gas supply chain in Brazil, *International Journal of Production Research* 59(10): 2960–2976.

Ghadge A, Er M, Ivanov D, Chaudhuri A (2022) Visualisation of ripple effect in supply chains under long-term simultaneous disruptions: A system dynamics approach, *International Journal of Production Economics* 60(20): 6173–6186.

Ghosh V, Paul A, Zhu L (2022) Stocking under random demand and product variety: Exact models and heuristics, *Production & Operations Management* 31(3): 1006–1032.

Hasanly A, Khajeh Talkhoncheh M, Karimi Alavijeh M (2018). Techno-economic assessment of bioethanol production from wheat straw: A case study of Iran, *Clean Technologies & Environmental Policy* 20(2): 357–377.

Hong J, Alzaman C, Diabat A, Bulgak A (2019) Sustainability dimensions and PM2.5 in supply chain logistics, *Annals of Operations Research* 275(2): 339–366.

Liao H, Shen N, Wang Y (2020) Design and realization of an efficient environmental assessment method for 3R systems: A case study on engine remanufacturing, *International Journal of Production Research* 58(19): 5980–6003.

Qazi A, Simselder MCE, Formaneck S (2023) Supply chain risk network value at risk assessment using Bayesian belief networks and Monte Carlo simulation, *Annals of Operations Research* 322(1): 241–272.

Sterman J (2010) *Business Dynamics: Systems Thinking and Modelling for Complex World*. New York: Irwin/McGraw-Hill.

Wang Y, Huang P, Wang L, Wang P, Wei K, Zhang Z, Yan B (2019) An improved ENSO ensemble forecasting strategy based on multiple coupled model initialization parameters, *Journal of Advances in Modeling Earth Systems* 11(6): 2868–2878.

6

Regression

In Chapter 3 we discussed correlation – a measure of strength of the association between numerical variables. Decision-makers are often interested in predicting the value of a dependent variable from the value of one or more independent, or explanatory, variables. For example, predicting the demand for a product (the dependent variable) as a function of price and advertising (the independent variables) is important in many retail operations. This model might be used to help determine appropriate values of price and advertising. Regression analysis is a tool for building statistical models that characterize relationships between a dependent variable and one or more independent variables, all of which are numerical.

Two broad categories of regression models are used often in business settings: (1) regression models of time series data (in which the independent variable is time and the focus is on predicting the future) and (2) regression models of cross-sectional data (data collected for a single time period, or when time is not a factor, and which usually involves multiple independent variables). Time series regression is an important tool in forecasting. But time series can have nonlinear forms, such as quadratic or more complicated relationships over time. In those cases, autoregressive moving average (ARIMA) and other modeling tools can sometimes be more effective.

In this chapter we describe how to develop and analyze both single and multiple regression models. Single variable regression (ordinary least squares, OLS) will be presented in R followed by ARIMA modeling in SAS and in R. Our principal focus is to gain a basic understanding of the assumptions of regression models, statistical issues associated with interpreting regression results, and practical issues in using regression as a tool for making and evaluating decisions. We will investigate

- Simple linear regression, focusing on the development of the model, understanding the significance of regression and the use of hypothesis tests, and interpreting key statistics such as the coefficient of determination and standard error of the estimate,
- ARIMA modeling with both SAS (which has interesting graphics useful in aiding modeling and output) as well as R,

DOI: 10.1201/9781003562399-6

- Multiple linear regression, including understanding how to select variables in models and evaluate goodness of fit and incorporating ordinal and nominal variables in the models, and
- Developing regression models to capture nonlinearities in data.

We will use data on the S&P 500 for single regression forecasting and the file InsuranceExpenseKaggle.csv for multiple regression.

6.1 Simple Linear Regression

We first discuss the case of a single independent variable. The relationship between two variables may be linear, one of many types of nonlinear forms, or there may be no relationship at all. To develop a regression model, we must first specify the type of function that best describes the data. This is important, because using a linear model for data that is clearly nonlinear, for instance, would probably lead to poor business decisions and results.

6.1.1 Least Squares Estimation

Because we will only have sample data available, we must estimate the parameter values that define the dependency relationship, i.e., of β_0 as the model intercept and β_1 as the parameter coefficient for the independent variable, in the case of linear regression. Let us call these estimates b_0 and b_1, respectively. Then $\hat{Y} = b_0 + b_1 X$ represents the estimate of the population mean for any value of independent variable X within the range of the sample data. The best estimates of the population parameters are typically determined by choosing estimates b_0 and b_1 to minimize the function of the errors associated with all the data. The most common approach for doing this is called **least-squares regression**, which minimizes the sum of squares of the errors:

$$\sum_{i=1}^{n}\left(Y_i - \left[b_0 + b_i X_i\right]\right)^2$$

Note that X_i and Y_i are the known observations and that b_0 and b_I are unknowns. Using calculus, we can show that the solution that minimizes this function is:

$$b_i = \frac{\sum_{i=1}^{n} X_i Y_i - n\bar{X}\bar{Y}}{\sum_{i=1}^{n} X_i^2 - n\bar{X}^2}$$

$$b_0 = Y - b_i X$$

Although these calculations appear to be somewhat complicated, they can easily be performed on an Excel spreadsheet or in other software, e.g., R. Fortunately, Excel has built-in capabilities to do this.

6.1.2 Coefficient of Determination and Correlation Coefficient

These sums of squares provide information about the strength of the regression relationship. Specifically, the ratio SSR/SST gives the proportion of variation that is explained by the independent variable of the regression model and is called the **coefficient of determination, R^2**. The value of R^2 will be between 0 and 1. A value of 1 indicates a perfect fit, while a value of 0 indicates that no relationship exists. The value of R^2 means that R^2 of the variation in the dependent variable is explained by regression (i.e., the independent variable, time). The remaining variation is due to other factors that were not included in the model. Although we would like high values of R^2, it is difficult to specify a "good" value that signifies a strong relationship, as this depends on the application. For example, in marketing research studies, an R^2 of 0.6 or more is considered very good, whereas in many social science applications even smaller values are considered acceptable.

The square root of the coefficient of determination is the sample correlation coefficient, R. Values of R range from −1 to 1, where the sign is determined by the sign of the slope of the regression line. The correlation table consists of these R measures. A correlation coefficient of R = 1 indicates perfect positive correlation; that is, as the independent variable increases, the dependent variable does also; R = −1 indicates perfect negative correlation – as X increases, Y decreases. As with R^2, a value of R = 0 indicates no correlation. Because R^2 measures the actual proportion of the variation explained by regression, it is usually easier to interpret than R.

6.1.3 Standard Error of the Estimate and Confidence Bands

Sums of squares divided by their appropriate degrees of freedom provide estimates of variances, which we call mean squares. The sum of squares SSE has two degrees of freedom because we estimated two parameters of the regression line, the slope and intercept. Thus, MSE = SSE/(n–2) is the variance of the errors about the regression line. The square root of MSE is called standard error of the estimate, $\mathbf{S_{YX}}$, and is simply the standard deviation of the errors about the regression line. This is computed as:

$$S_{YX} = \sqrt{\frac{SSE}{n-2}}$$

Intuitively, S_{YX} measures the spread of the data about the regression line. If the data are clustered close to the regression line, then the standard error will be small.

The standard error can be used to develop confidence bands around the regression line that are analogous to confidence intervals for point estimates. You might think of a confidence band as a collection of confidence intervals, one for each estimate of the dependent variable. A 100(1–α)% confidence band around the regression line is:

$$\hat{Y}_i \pm t_{\alpha/2,n-2} S_{YX} \sqrt{h_i}$$

where $\hat{Y}_i = b_0 + b_1 X_i$ is the predicted mean of Y for a given value, X_i of the independent variable, and:

$$h_i = \frac{1}{n} + \frac{(X_i - \bar{X})^2}{\sum_i^n (X_i - \bar{X})^2}$$

From this formula you can see that h_i varies with the value of X; in particular, h_i will be larger for values of X farther away from its mean, implying that the confidence intervals will be wider and that more uncertainty exists about the true population mean.

6.1.4 S&P Time Series Data

This dataset gives the beginning of the quarter price for the S&P 500 index, a widely reported and observed measure of blue-chip investments in the US. Using the S&P.csv datafile, the type of mathematical form (linear, quadratic, more complex) can usually be seen in a scatter diagram, and we always recommend that you create one first to gain some understanding of the nature of any potential relationship (see Figure 6.1).

The simplest type of regression model is a linear relationship involving one independent variable, X, and one dependent variable, Y, as illustrated in Figure 6.1 We assume that for any value of X, we have a distribution of values of Y – a population – whose mean is $\beta_0 + \beta_1$ X. Therefore, $Y = \beta_0 + \beta_1 X$ represents the equation of a straight line (the regression line) through the means of these populations. Here, β_0 is the intercept of the line, that is, the value of Y when X = 0; and β_1 is the slope, which measures the change in Y as X is increased by 1. However, individual data points will not lie on this line; therefore, we must include an error term to characterize the individual data points in the population. The mathematical form of this model is

$$Y = \beta_0 + \beta_1 X + \varepsilon$$

FIGURE 6.1
Scatter Diagram of S&P Price versus Time (time 1–264, the end of 2022).

where ε is an error term. For example, suppose that one observation of the ith dependent variable is Y_i when the value of the independent variable is X_i. Then the error associated with this data point is

$$\varepsilon_i = Y_i - \beta_0 - \beta_1 X_i$$

The error is positive if the data point lies above the line, and negative if it is below. Another way of looking at this is that the error term ε_i represents the difference between the mean value of the dependent variable $(Y_i = \beta_0 + \beta_1 X_i)$ and the actual value, Y_I, for a given value of X.

The R syntax to obtain a simple ordinary least squares regression of the S&P 500 time series 2001 through 2023 using monthly data is:

```
#TS Analysis and S&P data
install.packages(c("xts",      "forecast",      "tseries"),
repos = "http://cran.us.r-project.org")
library(xts)
library(forecast)
library(tseries)
SandP=read.csv("S&PData.csv")
SandP$Price=as.numeric(SandP$Price)
SandP_ts=ts(SandP$Price, start = c(2001,1), frequency = 12)
plot.ts(SandP_ts)
```

The output of the ordinary least squares regression for S&P 500 data is:

```
#model0- simple linear regression
m0= lm(Price~Time, data=SandP)
summary(m0)

##
## Call:
## lm(formula = Price ~ Time, data = SandP)
##
## Residuals:
##      Min      1Q  Median      3Q     Max
## -766.66 -467.47  -29.59  290.85 1309.35
##
## Coefficients:
##             Estimate Std. Error t value Pr(>|t|)
## (Intercept) 329.1220    60.3694   5.452 1.11e-07 ***
## Time         12.0674     0.3765  32.055  < 2e-16 ***
## ---
## Signif. codes:  0 '***' 0.001 '**' 0.01 '*' 0.05 '.'
0.1 ' ' 1
##
## Residual standard error: 501 on 275 degrees of freedom
## Multiple R-squared:  0.7889, Adjusted R-
squared:  0.7881
## F-statistic:  1027 on 1 and 275 DF,  p-value: <
2.2e-16

newdata=data.frame(Time=c(279:290))
forecasts=predict(m0, newdata = newdata)
forecasts
```

The R-squared of 0.7889 indicates a clear trend over time, but over 20 percent of the variation in the S&P is due to other causes.

The forecasts for 2024 are given in Table 6.1.

These forecasts are very low relative to actual values as shown in Figure 6.2.

The Akaike information criterion (AIC) measure for this regression is 3,256 which can be used to compare fit with other models such as ARIMA. The reason the forecast of the OLS regression model is low is explained by looking at Figure 6.1, which shows a slightly exponential increase in the S&P 500 over the time period. With this data it might have worked better to use a shorter period for building the model. The trend over the 23-year period was

TABLE 6.1

Forecasts

Month	Time	Price	Actual
Jan-24	279	3695.925	4846
Feb-24	280	3707.992	5096
Mar-24	281	3720.060	5254
Apr-24	282	3732.127	5036
May-24	283	3744.194	5278
Jun-24	284	3756.262	5460
Jul-24	285	3768.329	5522
Aug-24	286	3780.397	5648
Sep-24	287	3792.464	5762
Oct-24	288	3804.531	5705
Nov-24	290	3816.599	6032
Dec-24	291	3828.666	5882

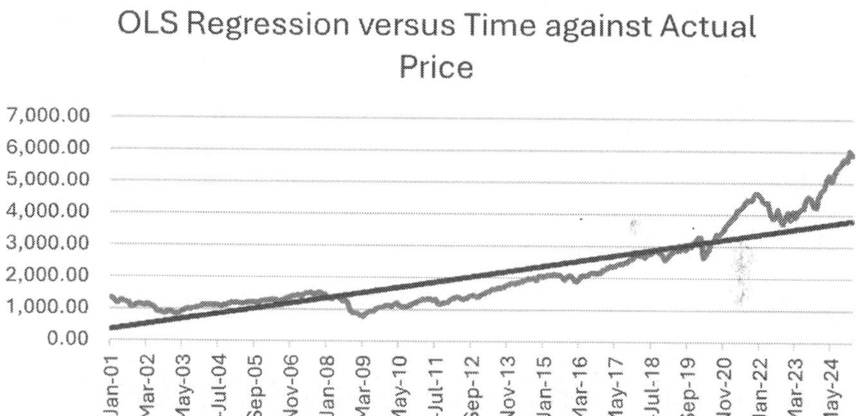

FIGURE 6.2
Actual versus OLS Regression.

an increase in the S&P 500 of 12.067 points per month. But the rate of increase got larger with time, leading to a gap between the actual and observed values.

Nonetheless, the model fit well, with an R^2 of 0.7889, implying that time explained about 79 percent of the increase in the index. Here adjusted R^2 means nothing as there is only one independent variable (Time). The *p*-value of the independent variable is based on *t*-score which was 32.055, very high. The inference is that there is 0.000 probability that time has nothing to do with the S&P 500 index.

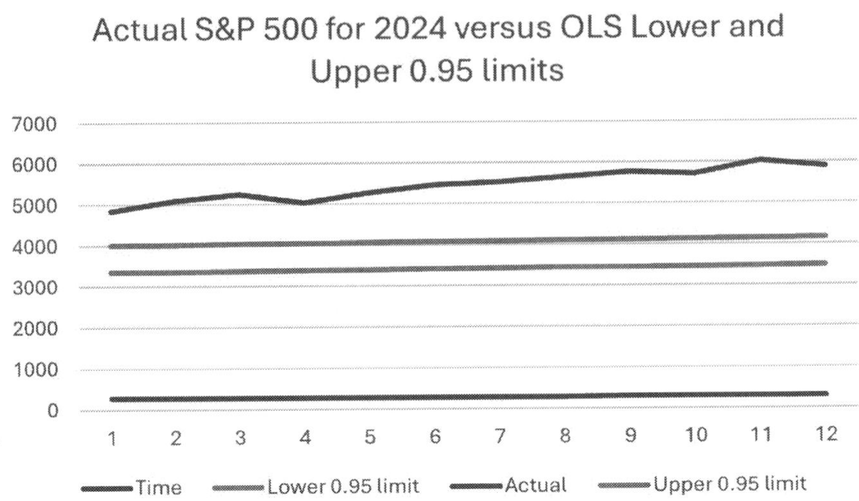

FIGURE 6.3
S&P 500 Forecast with Upper and Lower Limits.

Figure 6.3 shows that the 95 percent limits do not include the actual data, indicating that despite explaining about 76 percent of variation, there is a major bias in the forecast due to the nonlinear nature of the dataset.

6.2 ARIMA

Regression analysis is predicated on some key assumptions about the data. Clearly the first assumption is linearity. This is usually checked by examining a scatter diagram of the data or examining the residual plot. If the model is appropriate, then the residuals should appear to be randomly scattered about zero, with no apparent pattern. If the residuals exhibit some well-defined pattern, such as a linear trend, a U-shape, and so on, then there is good evidence that some other functional form might better fit the data.

The next key assumption is that the errors for each individual value of X are normally distributed with a mean of 0 and a constant variance. This can be verified by examining a histogram of the residuals associated with each value of the independent variable, and inspecting for a bell-shaped distribution, or using more formal goodness-of-fit tests as described in Chapter 4.

The third assumption is **homoscedasticity**, which means that the variation about the regression line is constant for all values of the independent variable. This can also be evaluated by plotting the residuals and looking for large differences in the variances at different values of the independent variable.

Finally, residuals should be independent for each value of the independent variable. This is an important assumption when time is the independent variable. If successive observations appear to be correlated, for example, by becoming larger over time or exhibiting a cyclical type of pattern, then this assumption is violated. Correlation among successive observations over time is called **autocorrelation** and can be identified by residual plots having clusters of residuals with the same sign. The pattern of residuals will have a cyclical pattern above and below the regression line, might lead us to suspect autocorrelation, which can be evaluated more formally using a statistical test based on the **Durbin-Watson statistic**. The Durbin-Watson statistic is

$$D = \frac{\sum_{i=2}^{n} (e_i - e_{i-1})^2}{\sum_{i=1}^{n} e_1^2}$$

This is a ratio of the squared differences in successive residuals to the sum of the squares of all residuals. When successive residuals are positively autocorrelated, D will approach 0. For our example, the value of D is computed to be 0.927. Critical values of the statistic have been tabulated based on the sample size and number of independent variables. For simple linear regression with $n = 20$, the critical value is 1.20, below which we would conclude that positive autocorrelation exists. That is the case in our example, suggesting that least squares regression is inappropriate for these data and that our statistical inferences may not be valid. Thus, before drawing inferences about regression models and performing hypothesis tests, these assumptions should be checked.

ARIMA models (AutoRegressive Integrated Moving Average) provide a means to fit a time series model incorporating cycles and seasonality in addition to trend. Box-Jenkins models are of this class. ARIMA models have up to three parameters: autocorrelation terms, differencing terms, and moving average terms. These will be discussed in the Box-Jenkins section. Exponential smoothing is a special case of the Box-Jenkins technique. However, while exponential smoothing is very computationally efficient, ARIMA models require large amounts of computation time. Furthermore, because so many parameters are used, larger amounts of data are required for reliable results. ARIMA works very well when time series data has high degrees of autocorrelation, but rather poorly when this condition does not exist. It usually is a good idea to test for autocorrelation and compare the fit of the ARIMA model with linear regression against time or some other forecasting model. A regression of the variable to be forecasted versus time is a special case of ARIMA (0 autocorrelation terms, 0 moving average terms).

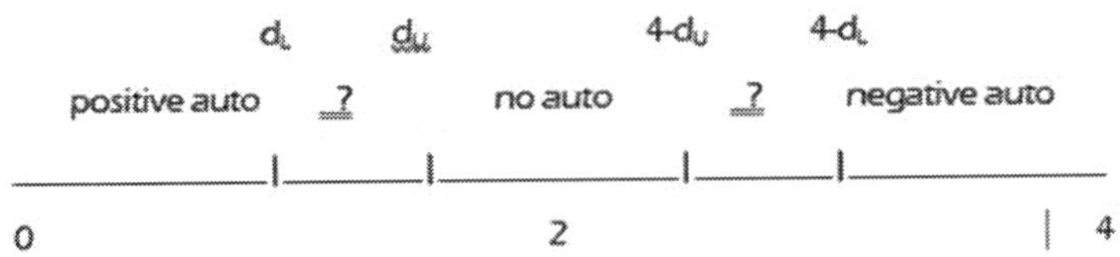

FIGURE 6.4
Durbin-Watson Scale.

The Durbin-Watson test provides an estimate of autocorrelation in a regression model. The null hypothesis of this test is that there is no autocorrelation in a regression model. Durbin-Watson statistics can range between 0 and 4. The ideal measure indicating no autocorrelation is 2. Values for lower and upper Durbin-Watson limits at the 0.95 confidence level; d_L is the lower limit, and d_U is the upper limit as available in statistics books. You need a computer regression package to obtain *d*, the estimate of first order autocorrelation. Then obtain d_L and d_U from a Durbin-Watson table (**k′** is the number of non-intercept independent variables, and *n* is the number of observations). Figure 6.4 gives limits and a plot of the Durbin-Watson scale:

To test for positive autocorrelation (ε_i is directly related to ε_j)

if *d* is less the d_L rejects the null (conclude there is positive autocorrelation)
if *d* is greater than d_L but less than d_U there is no conclusion relative to positive autocorrelation
if *d* is greater than d_U accept the null (conclude no positive autocorrelation exists)

To test for negative autocorrelation (ε_i is inversely related to ε_j)

if *d* is less than $4–d_U$ accept the null (conclude no negative autocorrelation exists)
if *d* is greater than $4–d_U$ and less than $4–d_L$ there is no conclusion relative to negative autocorrelation
if *d* is greater than $4–d_L$ reject the null (conclude the existence of negative autocorrelation)

There is a continuum for the evaluation of *d*.

If autocorrelation exists in a regression against TIME ($Y = f\{\text{time}\}$), this feature can be utilized to improve the forecast through a Box-Jenkins model. Second-stage least squares is an alternative approach, which runs the OLS regression, identifies autocorrelation, then adjusts the data to eliminate the autocorrelation, regresses on the data, and replaces the autocorrelation. As you can see, second-stage least squares is rather involved. (The SAS syntax for second-stage least squares regression is given at the end of the chapter.)

To summarize autocorrelation, the error terms are no longer independent. One approach (Box-Jenkins) is to utilize this error dependence to develop a better forecast. The other approach (second-stage least squares) is to wash the error dependence away.

6.3 Box-Jenkins Models

Box-Jenkins models were designed for time series with:

No trend

Constant variability

Stable correlations over time

Box-Jenkins models have a great deal of flexibility. You must specify three terms:

1. **P** – the number of autocorrelation terms
2. **D** – the number of differencing elements
3. **Q** – the number of moving average terms

The **P** term is what makes a Box-Jenkins model work, taking advantage of the existence of strong autocorrelation in the regression model $Y=f(\text{time})$.

The **D** term can sometimes be used to eliminate trends. **D** of 1 will work well if your data has a constant trend (it's linear). **D** of 2 or 3 might help if you have more complex trends. Going beyond a **D** value of 3 is beyond the scope of this course. If there is no trend to begin with, **D** of 0 works well.

The model should also have constant variability. If there are regular cycles in the data, moving average terms equal to the number of observations in the cycle can eliminate these. Looking at the plot of the data is the best way to detect cyclical data. One easily recognized cycle is seasonal data. If you have monthly data, a moving average term **Q** of 12 would be in order. If you have quarterly data, **Q** of 4 should help. If there is no regular pattern, **Q** of 0 will probably be as good as any.

D and **Q** terms are used primarily to stabilize the data. **P** is the term which takes advantage of autocorrelation. The precise number of appropriate autocorrelation terms (**P**) to use can be obtained from the computer package. **P** is the number of terms significantly different from 0. Significance is a matter of judgment. Since Box-Jenkins models are often exploratory, you will want to try more than one model anyway, to seek the best fit (lowest mean square forecasting error).

Box-Jenkins models tend to be volatile. They are designed for datasets of at least 100 observations. You won't always have those many observations. We are looking at them as an alternative, especially when autocorrelation is present in a regression versus time. So, the idea is to compare different models and select the best one.

Box-Jenkins models require a computer package for support. There are a number of them available. Minitab and SAS are other sources in addition to R. Specific operating instructions require review of corresponding manuals. SAS requires less parameter settings but is a little more rigid. Minitab commands for Box-Jenkins are very easy, also given at the end of the chapter.

6.3.1 SAS ARIMA

An ARIMA model from SAS using the syntax:

```
proc sgplot data=A;
Scatter y=Price x=Time;
Run;
```

SAS ARIMA modeling yields Figure 6.5.

The data is monthly, implying a seasonality of 12. The plot in Figure 6.5 has a quadratic looking feature, so a difference of 2 would be useful to try. Using the input:

```
Proc arima data=A;
Identify variable = Price;
```

This provides initial analysis of the data relative to ARIMA, which is given in Table 6.2 and Figure 6.6.

The PACF plot indicates a strong partial autocorrelation for P of 1, with a slight positive autocorrelation for P of 2. We will start with PDQ of $p = 2$, $d = 2$, and $q = 12$.

```
Proc arima data=A;
Identify variable = Price;
Run;
Identify var=Price(2);
```

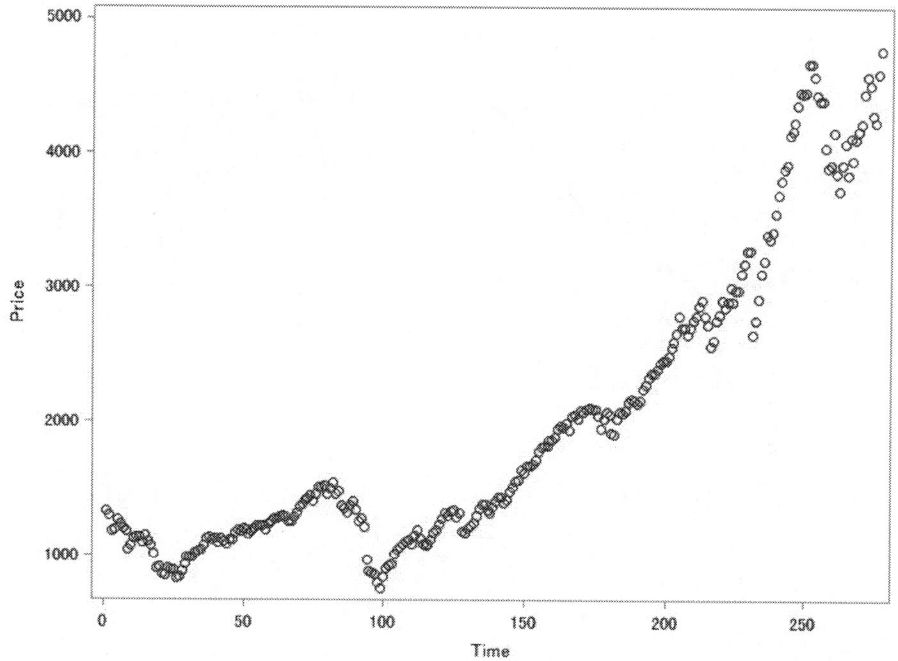

FIGURE 6.5
SAS ARIMA Initial Output.

TABLE 6.2
ARIMA Output for Initial Model: The ARIMA Procedure

Name of Variable = Price	
Mean of Working Series	2006.49
Standard Deviation	1086.421
Number of Observations	277

Autocorrelation Check for White Noise

To Lag	Chi-Square	DF	Pr > ChiSq	Autocorrelations					
6	1532.09	6	<.0001	0.984	0.969	0.958	0.946	0.931	0.915
12	2833.76	12	<.0001	0.900	0.887	0.872	0.859	0.847	0.834
18	3954.64	18	<.0001	0.823	0.810	0.800	0.791	0.780	0.766
24	4903.37	24	<.0001	0.755	0.744	0.732	0.717	0.702	0.687

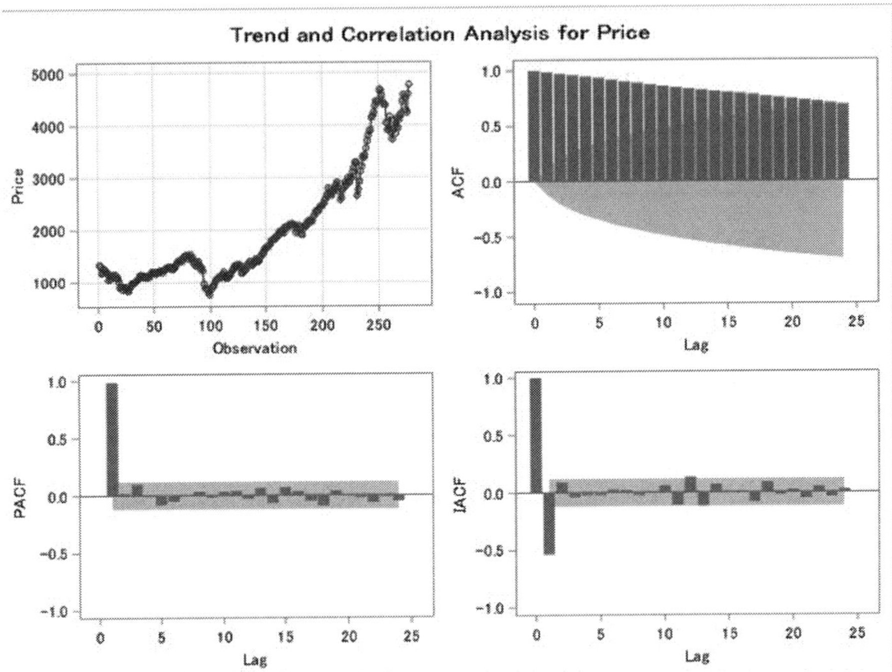

FIGURE 6.6
SAS Initial ARIMA Output.

```
Estimate p=2 q=12;
run;
forecast lead=12 id=Time out=results;
run;
```

Running this model gives output shown in Table 6.3 and Figure 6.7.

This yields output to include residual analysis in Figure 6.8 and Table 6.4 with forecast shown in Figure 6.9.

Various models were run exploring different combinations. The Akaike information criterions are compared.

P=1 D=1 Q=12	AIC 3264.092	
P=1 D=2 Q=4	AIC 3257.0	
P=1 D=2 Q=12	AIC=3264.1	
P=2 D=2 Q=1	AIC=3254.5	This combination suggested by Auto in R
P=2 D=2 Q=4	AIC=3240.7	this has the lowest AIC
P=2 D=2 Q=12	AIC=3256.2	This is the theoretically best model
OLS regression	AIC=3446.0	for comparison

TABLE 6.3

SAS (2,2,12) ARIMA Output for Model

Name of Variable = Price	
Period(s) of Differencing	2
Mean of Working Series	24.44756
Standard Deviation	130.2494
Number of Observations	275
Observation(s) Eliminated by Differencing	2

Autocorrelation Check for White Noise

To Lag	Chi-Square	DF	Pr > ChiSq	Autocorrelations					
6	74.68	6	<.0001	0.476	–0.083	–0.017	0.134	0.120	0.051
12	87.64	12	<.0001	0.104	0.117	–0.009	–0.103	–0.095	–0.032
18	103.61	18	<.0001	–0.093	–0.167	–0.035	0.088	0.024	–0.092
24	117.93	24	<.0001	–0.133	–0.126	–0.093	–0.056	–0.007	0.050

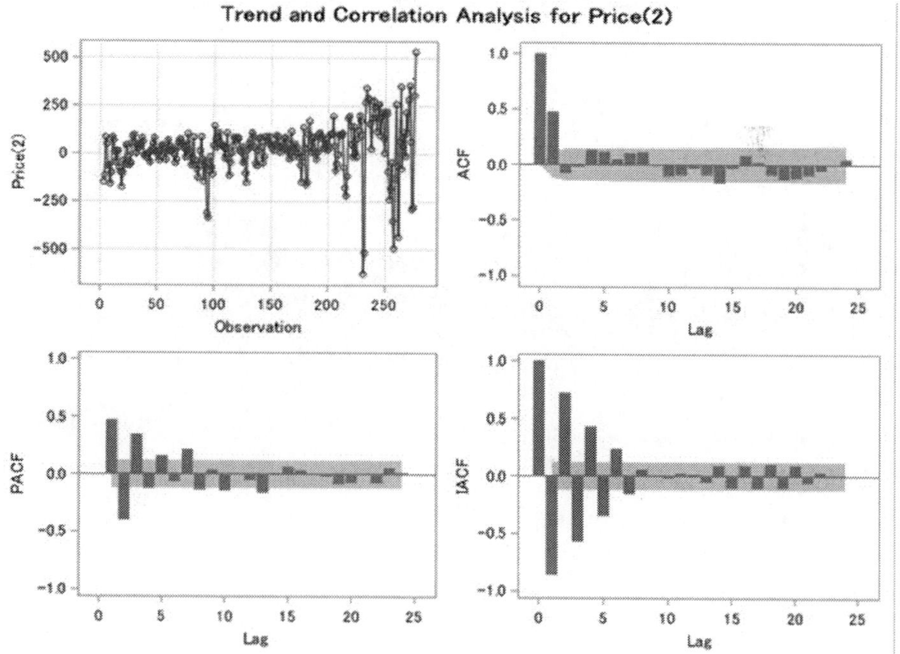

FIGURE 6.7
ARIMA Model (2,2,12) Output.

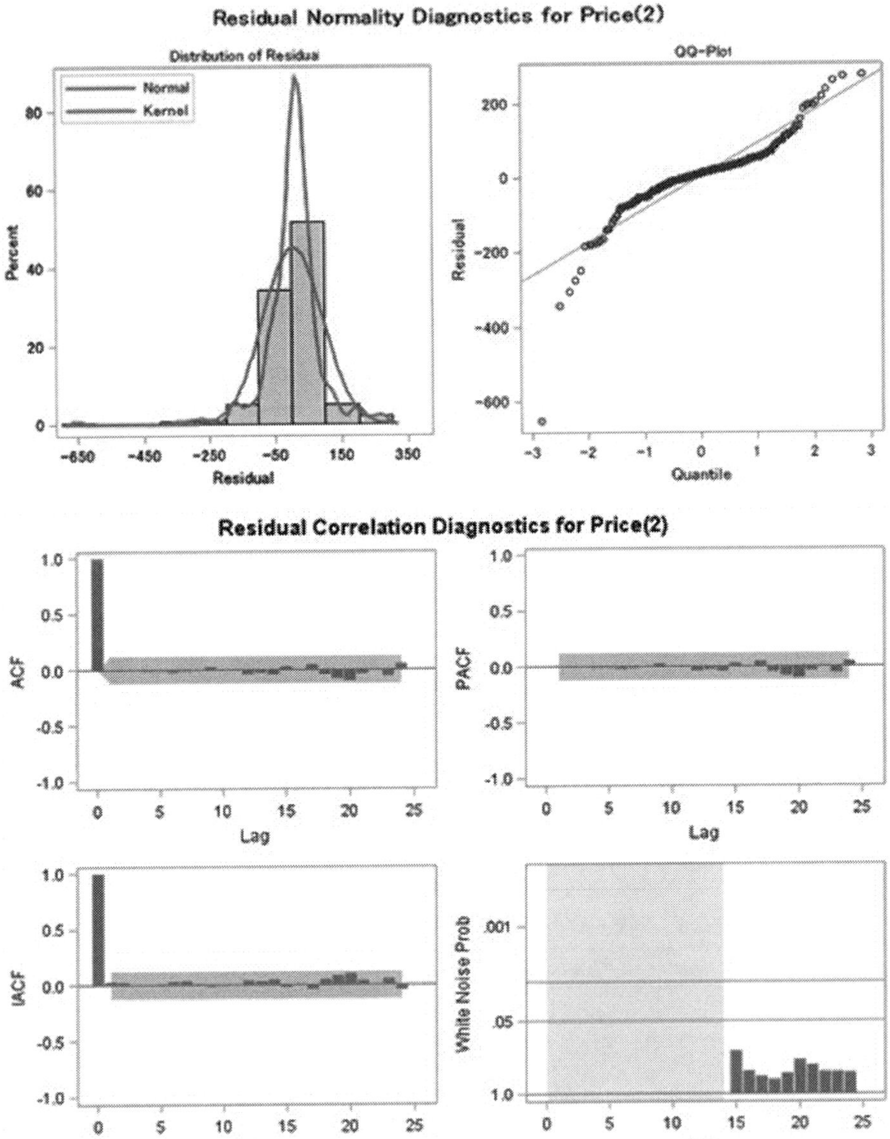

FIGURE 6.8
ARIMA (2,2,12) Residuals.

TABLE 6.4
SAS ARIMA (2,2,12) Model

Conditional Least Squares Estimation					
Parameter	**Estimate**	**Standard Error**	***t* Value**	**Approx Pr > |*t*|**	**Lag**
MU	16.49680	13.00217	1.27	0.2057	0
MA1,1	–1.38680	0.13090	–10.59	<.0001	1
MA1,2	–1.11127	0.20264	–5.48	<.0001	2
MA1,3	–0.67947	0.17265	–3.94	0.0001	3
MA1,4	–0.006889	0.13170	–0.05	0.9583	4
MA1,5	–0.14812	0.12895	–1.15	0.2517	5
MA1,6	–0.32168	0.13069	–2.46	0.0145	6
MA1,7	–0.44158	0.13341	–3.31	0.0011	7
MA1,8	–0.44091	0.13458	–3.28	0.0012	8
MA1,9	–0.23290	0.13689	–1.70	0.0901	9
MA1,10	–0.01028	0.13235	–0.08	0.9382	10
MA1,11	0.24049	0.11488	2.09	0.0373	11
MA1,12	0.19447	0.06919	2.81	0.0053	12
AR1,1	–0.32510	0.12195	–2.67	0.0082	1
AR1,2	–0.76375	0.11883	–6.43	<.0001	2

Constant Estimate	34.45934
Variance Estimate	7706.479
Std Error Estimate	87.78655
AIC	3256.191

The theoretical model of ARIMA(2,2,12) didn't fit as well as ARIMA(2,2,4). The first tries monthly seasonality, which seems appropriate for monthly data. But possibly there is some bias quarterly which might help explain why the ARIMA(2,2,4) model had the best AIC (along with being simpler, with 8 less estimated parameters).

We will explore P=2 D=2 Q=12 and P=2 D=2 Q=4 with R.

6.3.2 ARIMA Model and Output in R

There are several packages and functions in R that also support ARIMA analysis, with the added advantage of applying a machine learning tool to estimate P, D and Q values. We will compare these values used in the SAS output with the R output.

To run ARIMA in R we first install and load access libraries xts, forecast, and tseries:

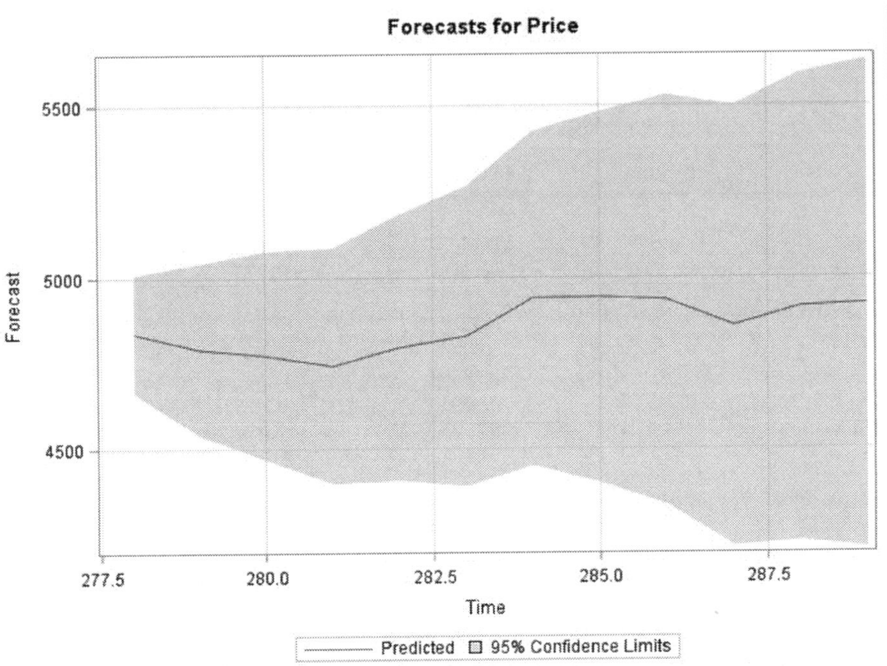

FIGURE 6.9
ARIMA Model Forecast with p=2, d=2, q=12.

```
#TS Analysis and S&P data
install.packages(c("xts", "forecast", "tseries"),
repos = "http://cran.us.r-project.org")
library(xts)
library(forecast)
library(tseries)
```

The next step is to link the data file S&PData.csv:

```
SandP=read.csv("S&PData.csv")
SandP$Price=as.numeric(SandP$Price)
SandP_ts=ts(SandP$Price, start = c(2001,1), frequency = 12)
plot.ts(SandP_ts)
```

Table 6.5 gives the forecasting model. We then plot this time series in Figure 6.10.

TABLE 6.5

ARIMA (2,2,12) Forecast

Model for Variable Price	
Estimated Mean	16.4968
Period(s) of Differencing	2
Autoregressive Factors	
Factor 1:	1 + 0.3251 B**(1) + 0.76375 B**(2)
Moving Average Factors	
Factor 1:	1 + 1.3868 B**(1) + 1.11127 B**(2) + 0.67947 B**(3) + 0.00689 B**(4) + 0.14812 B**(5) + 0.32168 B**(6) + 0.44158 B**(7) + 0.44091 B**(8) + 0.2329 B**(9) + 0.01028 B**(10) - 0.24049 B**(11) - 0.19447 B**(12)

Forecasts for Variable Price				
Obs	**Forecast**	**Std Error**	**95% Confidence Limits**	
278	4837.9110	87.7866	4665.8525	5009.9695
279	4791.6471	128.0360	4540.7011	5042.5931
280	4773.5150	155.3580	4469.0190	5078.0110
281	4742.9160	175.4844	4398.9729	5086.8591
282	4796.1914	198.2368	4407.6544	5184.7285
283	4827.9458	222.9820	4390.9092	5264.9824
284	4938.3368	248.8894	4450.5225	5426.1511
285	4941.0141	275.9784	4400.1063	5481.9219
286	4932.5748	304.9280	4334.9270	5530.2227
287	4857.3032	327.8285	4214.7712	5499.8352
288	4912.0028	347.6182	4230.6837	5593.3219
289	4921.0239	362.5480	4210.4430	5631.6049

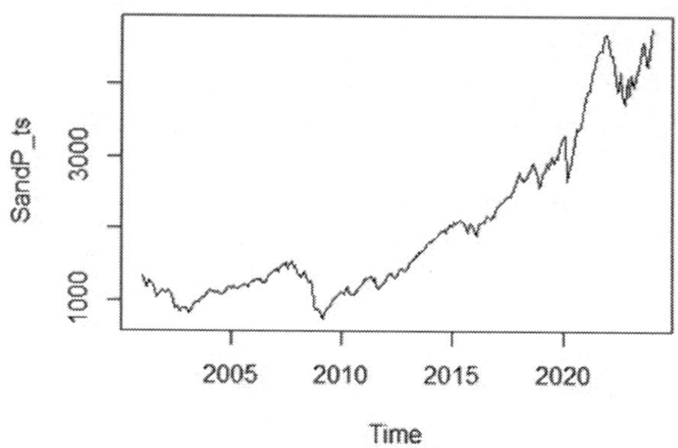

FIGURE 6.10
R S&P Trend Plot.

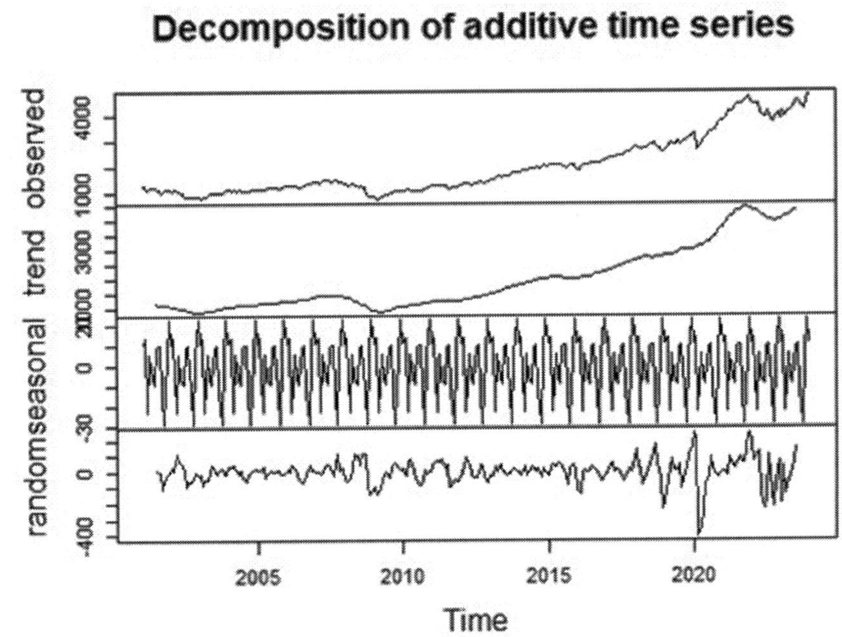

FIGURE 6.11
R Time Series Decomposition.

We can then decompose the S&P dataset for ARIMA analysis in Figure 6.11.

```
comp_SandP_ts=decompose(SandP_ts)
plot(comp_SandP_ts)
```

The trend in the second plot appears slightly quadratic, indicating a D value of 2. The third plot shows a seasonality of 12 (its monthly data), so a Q of 12 might be called for.

We can then run a stationary check using the Dickey-Fuller test, which has the null hypothesis that a unit root is present implying a tendency for the data to return to the mean:

```
#ADF stationary check
adf.test(SandP_ts)

##
## Augmented Dickey-Fuller Test
##
```

```
## data: SandP_ts
## Dickey-Fuller = -1.3751, Lag order = 6, p-value = 0.8393
## alternative hypothesis: stationary
```

Here the *p*-value of 0.8393 indicates rejection of the null hypothesis of no variance from the mean. We can next identify autocorrelation factors:

```
#KPSS stationary check
#install.packages("tseries")
#library(tseries)
#kpss.test(SandP_ts)
#Nonstationary time series - apply differencing
#SandP_ts_Station=diff(SandP_ts, differences=1)
#ndiffs(SandP_ts)
#plot(SandP_ts)
#kpss.test(SandP_ts)
acf(SandP_ts, main= "ACF for TS")
```

Figure 6.12 indicates that autocorrelation factors are present for both lags of 1 and 2, indicating that a P of 2 is useful to investigate.

Based on this exploratory process, we can try an ARIMA model with P=2, D=2, and Q=4:

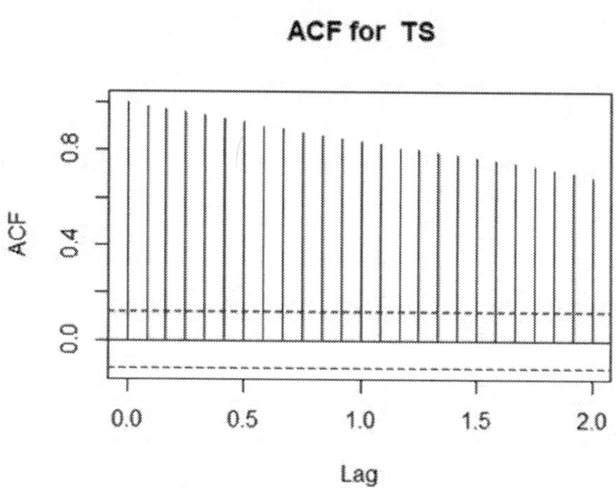

FIGURE 6.12
R Autocorrelation Factors for S&P 500 Data.

```
model1=Arima(SandP_ts, order=c(2,2,4))
summary(model1)
## Series: SandP_ts
## ARIMA(2,2,4)
```

This yields the output:

```
## Coefficients:
##              ar1      ar2      ma1     ma2      ma3     ma4
##          -0.0855  -0.8411   -0.829  0.6123  -0.7575  0.0061
## s.e.      0.1501   0.1227    0.164  0.1787   0.1447  0.0738
##
## sigma^2 = 7802:  log likelihood = -1621.26
## AIC=3256.52   AICc=3256.94   BIC=3281.84
##
## Training set error measures:
##                    ME      RMSE       MAE       MPE
## Training set 8.031335  87.04185  55.77981  0.4226735
MAPE      MASE
2.849328  0.2051796
##                     ACF1
## Training set -0.01015232
```

The ARIMA model's six coefficients are shown, each with their standard error. The coefficient for ar1 appears weak, while ar2 is strong, as are the first three moving average coefficients. The root mean square error of this model is 87.04, with an MAPE of 2.85. Figure 6.13 shows a plot of the residuals over the past dataset.

The Ljung-Box test is a test for the presence of autocorrelation in the model's residuals.

```
##
##  Ljung-Box test
##
## data:  Residuals from ARIMA(2,2,4)
## Q* = 27.506, df = 18, p-value = 0.06998
##
## Model df: 6.   Total lags used: 24
```

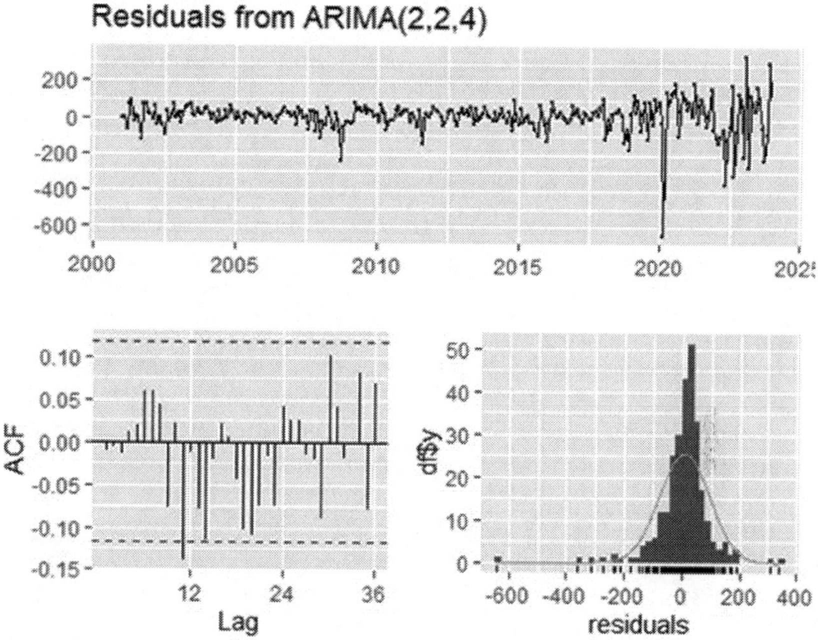

FIGURE 6.13
Plot of ARIMA(2,2,4) Model Residuals.

Here the p-value is 0.07, indicating a low probability that there is no autocorrelation among this model's residuals,. We can next forecast with the model. We set h at 12 to get 12 months' forecast.

```
#12 months forecast
autoplot(forecast(model1, h=12))
```

The output is shown in Figure 6.14, indicating that by the end of 2024 this model predicted the S&P would increase to around 5,000 (in fact by the end of November 2024 it reached 6,000).

We next try a simpler ARIMA(2,2,1) model, which is the output generated by R's auto ARIMA process:

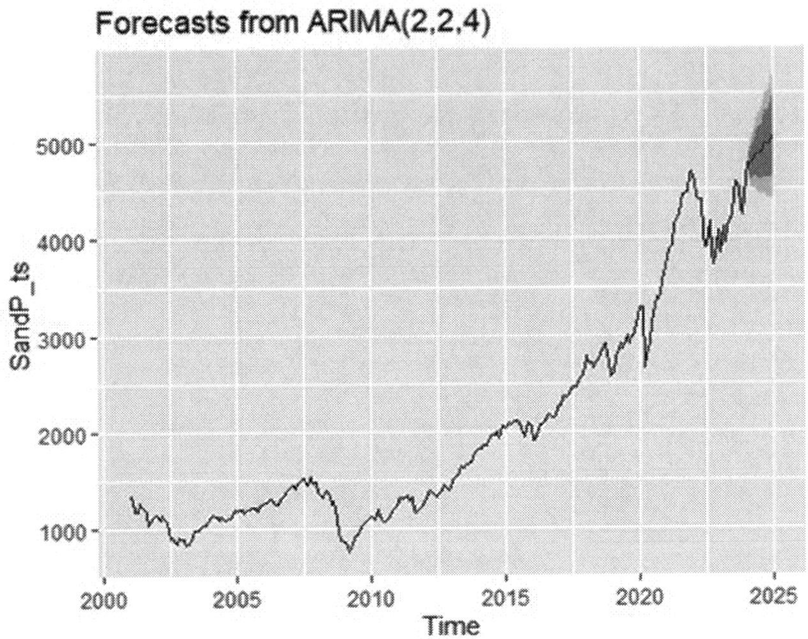

FIGURE 6.14
ARIMA(2,2,4) Forecasts.

```
model2=auto.arima(SandP_ts, seasonal = FALSE)
summary(model2)

## Series: SandP_ts
## ARIMA(2,2,1)
##
## Coefficients:
##          ar1      ar2      ma1
##       0.0696  -0.1140  -0.9828
## s.e.  0.0607   0.0623   0.0113
##
## sigma^2 = 7887:  log likelihood = -1624.23
## AIC=3256.47   AICc=3256.62   BIC=3270.94
##
## Training set error measures:
##                    ME     RMSE      MAE       MPE
## Training set 8.142522 88.00539 56.20367 0.4277564
MAPE      MASE
2.887117  0.2067387
##                     ACF1
## Training set -0.01539779
```

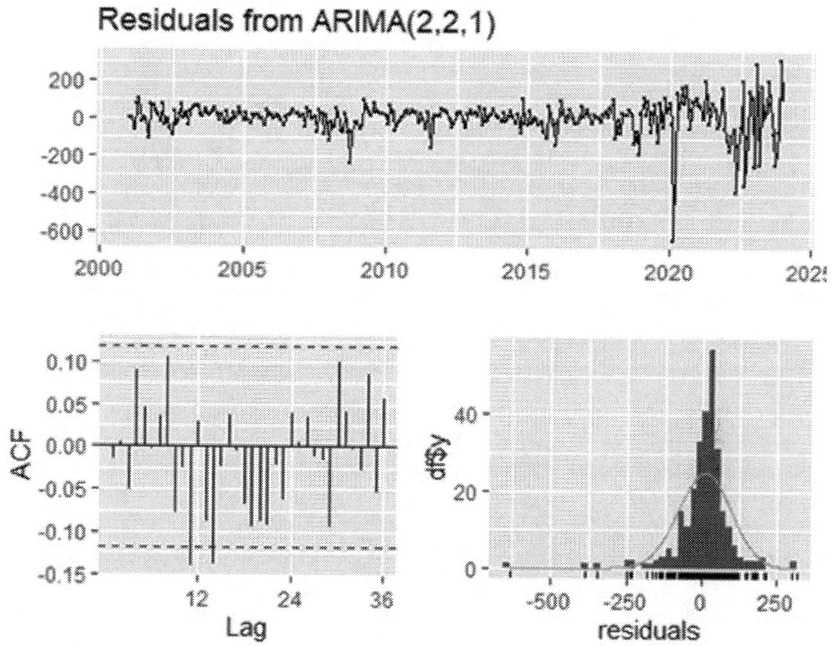

FIGURE 6.15
ARIMA(2,2,1) Residual Plot.

This yields the model shown here, with all three model coefficients significant at the 0.10 level and an Akaike information criterion measure is slightly (very slightly) better than that of the ARIMA(2,2,4) model built earlier. However, the root mean square error is slightly worse. Figure 6.15 plots the residuals of the ARIMA(2,2,1) model.

We also apply the Ljung-Box test:

```
##  Ljung-Box test
##
## data:  Residuals from ARIMA(2,2,1)
## Q* = 34.238, df = 21, p-value = 0.03417
##
## Model df: 3.   Total lags used: 24
```

This indicates a very complex model with lags of up to 34 periods. We can forecast with this model:

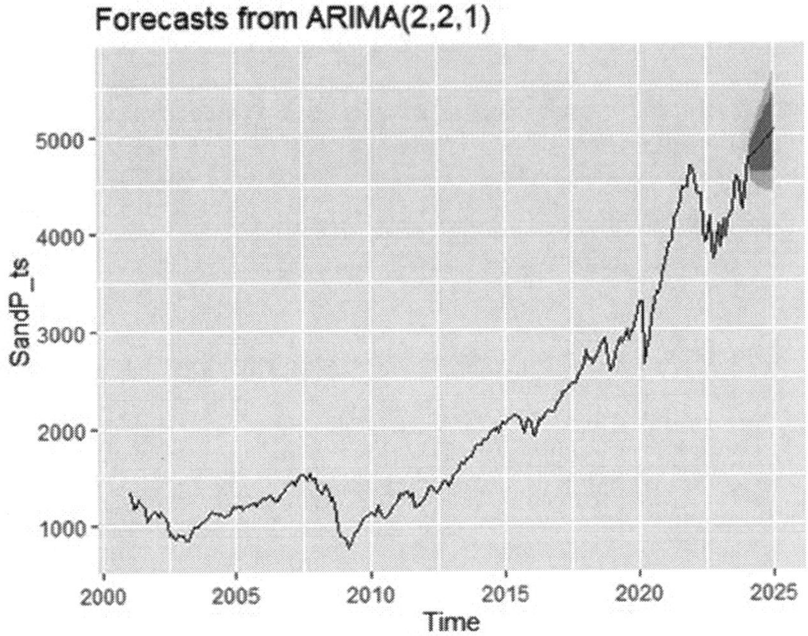

FIGURE 6.16
ARIMA(2,2,1) Forecast.

```
#12 months forecast
autoplot(forecast(model2, h=12))
```

This yields the plot shown in Figure 6.16.

This forecast predicts a slightly higher forecast in the vicinity of 5,200 by the end of 2024 (still quite below what happened).

ARIMA modeling can be quite effective if there is strong autocorrelation present (as indicated by Durbin-Watson or Ljung-Box measures). If there is a weak autocorrelation, other forecasting models usually do a lot better. But ARIMA provides a useful tool for data with certain features.

We next look at multiple regression.

6.4 Multiple Regression

6.4.1 Insurance Expense Dataset

This dataset (InsuranceExpenseKaggle.csv) is obtained from the public website Kaggle.com. It contains data on health insurance. There are 1,338 observations.

Region: Represents the geographical region of the insured individual's residence (not used here)

Age: Represents the age of the insured individual

Gender: Indicates the gender of the insured individual (male or female). (F=0, M=1)

BMI: Stands for Body Mass Index, a measure of body fat based on height and weight

Children: Denotes the number of children or dependents covered by the insurance policy

Smoker: Indicates whether the insured individual is a smoker or non-smoker (no=0; yes=1)

Expenses: Represents the medical expenses incurred by the insured individual

This dataset can be used for a number of things, to include predicting insurance charges based on the age, sex, and BMI of a customer. The data has been gathered from a variety of sources and contains information such as age, gender, region, and BMI values for each customer.

To make use of this dataset you will first need to understand the different variables present in it so you can understand which ones have an impact on predicting insurance charges. Age is expectedly one of the most important variables as younger or older customers may pay less or more respectively for their policies. Similarly, sex is also influential as traditionally gender roles dictate premiums with men paying more than women for the same coverage on many policies historically speaking. BMI should also be taken into account when making any predictions regarding insurance costs due to varying factors such as risk factors associated with obesity being taken into consideration by premium pricing decisions made by insurers. You can also consider the impact of the number of children covered, and whether the insured is a smoker or not. Once how all these elements influence pricing decisions is understood it is then time to explore potential predictive models that could accurately calculate an appropriate amount/estimation based on what you know about a customer's characteristics.

The first thing we can do is to run correlation analysis. Table 6.6 gives correlations for this data.

Table 6.6 shows that there is a very strong relationship between smoking and insurance expenses. There is also some correlation between age and a bit less with BMI. Gender and the number of children do not appear to matter as much. Note that all the correlations with expenses are positive, meaning that as the other variable increases, so do expenses. There is a negative correlation (very slight) between smoking and age – indicating a slight trend for older people not to smoke.

TABLE 6.6

Correlation of InsuranceExpenseKaggle.csv Data

	Age	Gender	Bmi	Children	Smoke	Expenses
Age	1.000					
Gender	–0.021	1.000				
Bmi	0.109	0.046	1.000			
Children	0.042	0.017	0.013	1.000		
Smoke	–0.025	0.076	0.004	0.008	1.000	
Expenses	0.299	0.057	0.199	0.068	0.787	1.000

The strongest regression will include all variables:

```
## Regression Modeling - Insurance Expense
install.packages("forecast", repos = "http://cran.us.r-project.org")

library(forecast)

data=read.csv("InsuranceExpenseKaggle.csv")
#variable conversion
data$smoke=as.factor(data$smoke)

#Model1 Using  inputs: age, F0M1, bmi, children, smoke
model1=lm(data$expenses~ age+ F0M1+ bmi+children+smoke,
data = data)
summary(model1)

##
## Call:
## lm(formula = data$expenses ~ age + F0M1 + bmi + children + smoke,
##     data = data)
##
## Residuals:
##      Min       1Q   Median       3Q      Max
## -11834.4  -2916.4   -990.6   1372.1  29554.5
##
## Coefficients:
##              Estimate Std. Error t value Pr(>|t|)
## (Intercept) -12055.16     951.23 -12.673  < 2e-16 ***
## age            257.72      11.90  21.650  < 2e-16 ***
## F0M11         -128.68     333.35  -0.386 0.699540
```

```
## bmi              322.45      27.42  11.761  < 2e-16 ***
## children         474.60     137.85   3.443 0.000593 ***
## smoke1         23822.31     412.51  57.750  < 2e-16 ***
## ---
## Signif. codes:  0 '***' 0.001 '**' 0.01 '*' 0.05 '.'
0.1 ' ' 1
##
## Residual standard error: 6070 on 1332 degrees of freedom
## Multiple R-squared:  0.7497, Adjusted R-squared:  0.7488
## F-statistic: 798.1 on 5 and 1332 DF,  p-value: < 2.2e-16
```

Note here that the *P*-value for gender (F0M1) is the only one that is not significant. From the correlation table in Table 6.6, there is little overlap in information across independent variables, which indicates little risk of multicollinearity. But the *P*-value for the full model indicates we could skip gender. Doing so yields the following output:

```
#Model 2 : input  age, bmi children smoke
model2=lm(data$expenses~     age+     bmi+children+smoke,
data = data)
summary(model2)
```

```
##
## Call:
## lm(formula = data$expenses ~ age + bmi + children +
smoke, data = data)
##
## Residuals:
##    Min     1Q Median     3Q    Max
## -11895  -2921   -985   1382  29499
##
## Coefficients:
##              Estimate Std. Error t value Pr(>|t|)
## (Intercept) -12105.48     941.95 -12.851  < 2e-16 ***
## age            257.83      11.90  21.674  < 2e-16 ***
## bmi            321.94      27.38  11.760  < 2e-16 ***
## children       473.69     137.79   3.438 0.000604 ***
## smoke1       23810.32     411.21  57.903  < 2e-16 ***
## ---
## Signif. codes:  0 '***' 0.001 '**' 0.01 '*' 0.05 '.'
0.1 ' ' 1
```

```
##
## Residual standard error: 6068 on 1333 degrees of freedom
## Multiple R-squared:  0.7497, Adjusted R-squared:  0.749
## F-statistic:   998.2 on 4 and 1333 DF,   p-value: <
2.2e-16
```

Note that this model has coefficients very close to those including gender. That is because there was little overlapping information across these independent variables, and the gender variable had little to do with experienced expenses. All four of the retained independent variables contributed to the predictive model, which would multiply case values for age, BMI, number of children, and smoking times the beta coefficients given.

6.4.2 Using Adjusted R^2 to Evaluate Fit

A useful way of examining the relative fit of different models is through the Adjusted R^2. Adding an independent variable to a regression model will always result in R^2 equal to or greater than the R^2 of the original model. This is true even when the new independent variable has little true relationship with the dependent variable. The Adjusted R^2, which we introduced earlier in the context of simple linear regression, reflects both the number of independent variables and the sample size. This helps to better understand the value of adding independent variables to the model. The Adjusted R^2 for a multiple linear regression models is computed as

$$\text{Adjusted } R^2 = 1 - \frac{\text{SSE}}{\text{SST}}\left(\frac{n-1}{n-k-1}\right)$$

where SSE = sum of squared errors (residuals)
SST = Total sum of squares
n= number of observations
k = number of independent variables

Essentially, the adjustment term $(n-1)/(n-k-1)$ represents the ratio of the degrees of freedom of SSE and SST. For the model with all four independent variables, we have:

$$\text{Adjusted } R^2 = 1 - \frac{49675447085}{196074221665} = 0.749$$

The Adjusted R^2 for the model without gender was also 0.749, indicating no major loss of information. We can demonstrate an extreme with the regression with the single independent variable of age:

```
#Model 3 : input using only Age
model3=lm(data$expenses~ age, data = data)
summary(model3)
```

```
##
## Call:
## lm(formula = data$expenses ~ age, data = data)
##
## Residuals:
##    Min     1Q Median     3Q    Max
##  -8059  -6671  -5939   5440  47829
##
## Coefficients:
##             Estimate Std. Error t value Pr(>|t|)
## (Intercept)   3165.9      937.1   3.378 0.000751 ***
## age            257.7       22.5  11.453  < 2e-16 ***
## ---
## Signif. codes:  0 '***' 0.001 '**' 0.01 '*' 0.05 '.'
0.1 ' ' 1
##
## Residual standard error: 11560 on 1336 degrees of
freedom
## Multiple R-squared:  0.08941,     Adjusted R-squared:
0.08872
## F-statistic: 131.2 on 1 and 1336 DF,  p-value: < 2.2e-16
```

For regression model with only age as the independent variable the Adjusted R^2 is 0.089. Adjusted R^2 provides a way to evaluate the effect of adding or removing variables to or from a model.

From a practical perspective, the independent variables selected should make some sense in attempting to explain the dependent variable (that is, you should have some reason to believe that changes in the independent variable will cause changes in the dependent variable even though causation cannot be proven statistically). From a statistical viewpoint, it is best to have as simple a model as possible. One approach to constructing a good model is to collect data on as many potential independent variables as possible (plus of course the dependent variable) and then try to eliminate those that have little effect on the dependent variable. This can be viewed as an effort to obtain *model efficiency*. The modern use of statistics, however, makes model efficiency much less important. Modern computers provide the ability to deal with masses of data easily, and tools such as data mining can seek models and variable associations that analysts might never uncover. However, techniques like data mining require a solid understanding of regression procedures to interpret computer output and to direct analysis. Analyzing the correlation matrix can help to identify appropriate independent variables.

Including multiple variables in the regression model may result in multicollinearity, a condition occurring when two or more independent variables in the same regression model contain high levels of the same information, thereby both explaining the variation in the dependent variable, and thus, are correlated with one another. Essentially, multicollinearity may be an issue when some set of independent variables predict each other better than they predict the dependent variable. Multicollinearity makes it difficult to distinguish the effects of these variables on the dependent variable. When multicollinearity is present, the β coefficients of the independent variables can be unstable, and even the signs of these β coefficients may change when different variables are included. Also, *p*-values can be inflated, resulting in the conclusion not to reject the null hypothesis for significance of regression when it should be rejected. Although a full discussion of multicollinearity is beyond the introductory scope of this book, it is important that you are aware of the problems that it may cause and seek expert advice when appropriate.

6.5 Conclusions

Regression is a powerful modeling tool to identify and understand relationships between independent and dependent variables. Regression is often applied to time series data as well as cross-sectional data. Time series often display nonlinear features, so ARIMA modeling can sometimes be more effective than OLS in those cases. Simple linear regression involves a single independent variable; multiple regression models involve several independent variables. Ordinal and nominal data can be included in regression models through coding.

Regression models have important business applications in predicting the results of decisions. Good regression models should be based on fundamental knowledge that the set of independent variables can cause changes in the dependent variable. This enables managers to focus on variables they can control, with confidence that their decisions will lead to the results anticipated. New techniques like data mining focus on the ability to predict rather than on statistical validity. However, a good fit says nothing about causation. Data may exhibit a statistical relationship, but in reality, variables may have little in common and may be due to a quirk of fate. As W. Edwards Deming often said, there is no knowledge without theory.

Estimating parameters for a linear regression model is done using the method of least squares. The significance of regression is evaluated by viewing regression as analysis of variance. Key statistics used to evaluate the results are the coefficient of determination, correlation coefficient, and the ANOVA F-test. In using regression, it is important to pay attention to the underlying assumptions used in the least squares method: normally distributed errors and constant variance. If these are violated, the statistical significance of the results may be questionable.

In multiple regression models, we must be cautious of multicollinearity, which can be avoided by careful selection of independent variables that do not include overlapping information content. However, the need for this depends upon the intended use of the model. The existence of multicollinearity does not impede the ability of the regression model to predict. It may make more sense to include additional independent variables that have overlapping information if they make sense from a theoretical perspective (that is, the user expects the independent variable in question to have an impact on the dependent variable).

7

Classification Tools

This chapter will look at classification tools, beginning with logistic regression, moving on to support vector machines, neural networks, and decision trees (to include random forests). We will demonstrate these methods with R programming, which also includes boosting. Classification uses a training dataset to identify classes or clusters, which then are used to categorize data. Typical applications include categorizing risk and return characteristics of investments, and credit risk of loan applicants. We will use a bankruptcy data file related to credit risk of organizations in China.

7.1 China Credit Risk Dataset

This dataset is taken from Kaggle.com. It has 32,581 observations, 7,108 of which defaulted on a loan, 25,473 did not. Some data is missing for employment length and loan interest rate. Table 7.1 describes variables (attributes). The site is openly available at:

www.kaggle.com/datasets/laotse/credit-risk-dataset

We will omit the categorical variables home ownership and loan intent as well as the variables employment length and loan interest rate due to missing observations. The target variable is Loan Status. The correlation among these variables is given in Table 7.2.

7.2 Logistic Regression

Some data of interest in a regression study may be ordinal or nominal. For instance, in our Chinese credit risk model, sex and college degree are binary. In the loan application data, the outcome is binary, while credit rating is ordinal. Since regression analysis requires numerical data, we included them by *coding* the variables. Here, each of these variables is dichotomous;

DOI: 10.1201/9781003562399-7

TABLE 7.1

Attributes in Credit Risk Dataset

Variable	Description	Type	Min	Max
person_age	Borrower age	Number	20	144
person_income	Borrower annual income	Number	4000	6000000
person_home_ownership	Does borrower own their home	Category (3)		
person_emp_length	Length of employment in years	Number	0	123
loan_intent	Borrower reason for requesting loan	Category		
loan_grade	Categorization of loan risk	Category (6)		
loan_amount	Borrowed	Number	500	35000
loan_interest_rate	Percentage interest	Number	5.42	23.22
loan_status	0-OK; 1-Default	Binary	0	1
loan_percent_income		Number	0	0.83
cb_person_default_on_file	Defaulted in past?	Binary		
cb_person_credit_hist_length		Number	2	30

TABLE 7.2

Correlation for Credit Risk Dataset Correlation

	Age	Income	Grade	Amount	Percentage	Default	History
Age	1.000						
Income	0.173	1.000					
Grade	0.014	-0.001	1.000				
Amount	0.051	0.267	0.146	1.000			
Percentage	-0.042	-0.254	0.123	0.573	1.000		
Default	0.006	-0.004	0.537	0.039	0.036	1.000	
History	0.859	0.118	0.015	0.042	-0.032	0.005	1.000
Status	-0.022	-0.144	0.373	0.105	0.379	0.179	-0.016

therefore, we can code them as either 0 or 1 (as we did in the regression model for loan applicants). For example, a male is assigned a code of 0, while a female is assigned a code of 1. The employees with a college degree can be assigned a code of 1, and those without a degree a code of 0.

The purpose of logistic regression is to classify cases into the most likely category. Logistic regression provides a set of β parameters for the intercept (or intercepts in the case of ordinal data with more than two categories) and independent variables, which can be applied to a logistic function to estimate the probability of belonging to a specified output class. The formula for probability of acceptance of a case i to a stated class j is:

$$P_j = \frac{1}{1+e^{\left(-\beta_0-\sum \beta_i x_i\right)}}$$

where, β coefficients are obtained from logistic regression.
To run a logistic regression model in R:

```
# China Credit Risk Data Set
#Model1 - Logistic Regression
CCreditRisk  =  as.data.frame(read.csv("Chinacredit_risk_
  dataset.csv"))
#variable conversion-
CCreditRisk$loan_status=as.factor(CCreditRisk$loan_
  status)
```

First we partition the data holding out 20 percent for testing:

```
#data partitioning
set.seed(123)
training.index     =     sample(c(1:dim(CCreditRisk)[1]),
  dim(CCreditRisk)[1]*0.8)
training.data = CCreditRisk[training.index,]
test.data = CCreditRisk[-training.index,]
```

To run the generalized logistic regression model in R:

```
log_m1<-glm(loan_status  ~  person_age  +  person_income
  +  loan_grade+  loan_amnt+  loan_percent_income+  cb_
  person_default_on_file+cb_person_cred_hist_length,
  data = training.data, family = "binomial")
```

This yields:

```
summary(log_m1)

##
## Call:
## glm(formula = loan_status ~ person_age + person_income + loan_grade +
##     loan_amnt + loan_percent_income + cb_person_default_on_file +
##     cb_person_cred_hist_length, family = "binomial", data = training.
data)
```

```
summary(log_m1)

##
## Call:
## glm(formula = loan_status ~ person_age + person_income + loan_grade +
##     loan_amnt + loan_percent_income + cb_person_default_on_file +
##     cb_person_cred_hist_length, family = "binomial", data = training.
data)
##
## Coefficients:
##                  Estimate Std. Error z value Pr(>|z|)
## (Intercept)   -3.576e+00  1.404e-01 -25.470  < 2e-16
***
## person_age                 -8.033e-04  5.950e-03  -0.135     0.893
## person_income               6.654e-08  7.775e-07   0.086     0.932
## loan_gradeB                 4.949e-01  5.190e-02   9.537   < 2e-16
## loan_gradeC                 8.669e-01  6.103e-02  14.206   < 2e-16
## loan_gradeD                 2.850e+00  6.506e-02  43.810   < 2e-16
## loan_gradeE                 3.156e+00  9.742e-02  32.396   < 2e-16
## loan_gradeF                 3.699e+00  1.830e-01  20.212   < 2e-16
## loan_gradeG                 6.973e+00  1.022e+00   6.821 9.04e-12
## loan_amnt                  -1.052e-04  5.374e-06 -19.580   < 2e-16
## loan_percent_income         1.239e+01  3.027e-01  40.933   < 2e-16
## cb_person_default_on_fileY  2.523e-02  5.281e-02  0.478    0.633
## cb_person_cred_hist_length -2.321e-03  9.191e-03  -0.252     0.801
## Signif. codes:  0 '***' 0.001 '**' 0.01 '*' 0.05 '.' 0.1 ' ' 1
```

Odds ratios for each variable are available:

```
####ODDS RATIOS###
round(exp(coef(log_m1)),3)

##                (Intercept)                 person_age
##                      0.028                      0.999
##              person_income                loan_gradeB
##                      1.000                      1.640
##                loan_gradeC                loan_gradeD
##                      2.380                     17.294
##                loan_gradeE                loan_gradeF
##                     23.474                     40.413
##                loan_gradeG                  loan_amnt
##                   1067.653                      1.000
##        loan_percent_income cb_person_default_on_fileY
##                 240345.737                      1.026
## cb_person_cred_hist_length
##                      0.998
```

The coincidence matrix and fit statistics can be obtained as follows:

```
### PLOTING AUCs#####
install.packages(c("rpart", "rpart.plot", "partykit",
"caret", "pROC"), repos = "http://cran.us.r-project.org")

# call packages and Load

library(caret)

Credit.pred= factor(ifelse (Credit.pred >0.5, 1, 0))

confusionMatrix(data=Credit.pred, reference = test.
data$loan_status, mode = "everything", positive = "1")

## Confusion Matrix and Statistics
##
##           Reference
## Prediction    0    1
##          0 4789  749
##          1  286  693
##
##                Accuracy : 0.8412
##                  95% CI : (0.8321, 0.85)
##     No Information Rate : 0.7787
##     P-Value [Acc > NIR] : < 2.2e-16
##
##                   Kappa : 0.4793
##
##  Mcnemar's Test P-Value : < 2.2e-16
##
##             Sensitivity : 0.4806
##             Specificity : 0.9436
##          Pos Pred Value : 0.7079
##          Neg Pred Value : 0.8648
##               Precision : 0.7079
##                  Recall : 0.4806
##                      F1 : 0.5725
##              Prevalence : 0.2213
##          Detection Rate : 0.1063
##    Detection Prevalence : 0.1502
```

```
##        Balanced Accuracy : 0.7121
##
##         'Positive' Class : 1
##

library(rpart)
library(partykit)

library(caret)
library(pROC)

library(rpart.plot)

#test_roc = roc(valid.df$Fraud ~ fraud.pred, plot = TRUE,
print.auc = TRUE)
###Fix the Plot area to Square####
par(pty="s")
Credit.pred <- predict(log_m1, test.data, type = "response")
test_roc  =  roc(test.data$loan_status  ~  Credit.pred,
plot = TRUE, print.auc = TRUE, legacy.axes=TRUE, percent=
TRUE,  xlab="False  Positive  Rate",  ylab="True  Positive
Rate", lwd=4)

## Setting levels: control = 0, case = 1

## Setting direction: controls < cases

###Return back to Maximum Plot Area###

par(pty="m")
```

The area under the recall curve is given in Figure 7.1.
To predict some of the test data with this model:

```
Credit.pred <- predict(log_m1, test.data, type = "response")

# first 10 actual and predicted records
data.frame(actual        =        test.data$loan_status[1:10],
predicted = ifelse(Credit.pred[1:10]>0.5, 1,0))
```

This yields:

```
##     actual predicted
## 12       1         1
## 14       1         0
## 25       1         1
## 34       0         0
## 35       0         0
## 40       1         1
## 53       1         0
## 61       0         0
## 65       1         1
## 70       0         0
```

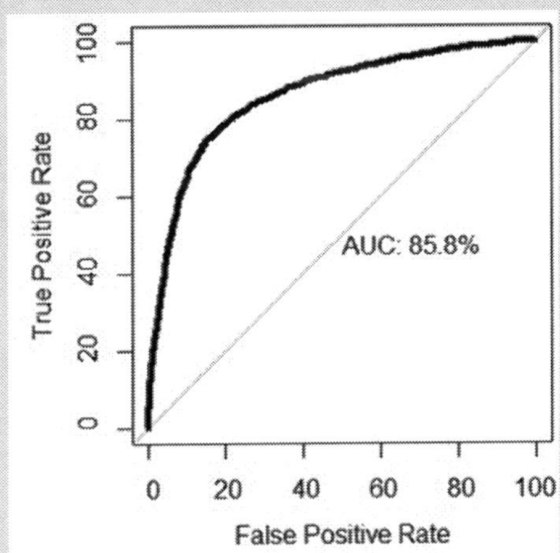

FIGURE 7.1
Area under the Curve for the Logistic Regression Model.

7.3 Support Vector Machines

Support vector machines (SVMs) are supervised learning methods that generate input-output mapping functions from a set of labeled training data. The mapping function can be either a classification function (used to categorize the input data) or a regression function (used for estimation of the desired output). For classification, nonlinear kernel functions are often used to transform the input data (inherently representing highly complex nonlinear relationships) to a high dimensional feature space in which the input data becomes more separable (i.e., linearly separable) compared to the original input space. Then, the maximum-margin hyperplanes are constructed to optimally separate the classes in the training data. Two parallel hyperplanes are constructed on each side of the hyperplane that separates the data by maximizing the distance between the two parallel hyperplanes. An assumption is made that the larger the margin or distance between these parallel hyperplanes the better the generalization error of the classifier will be.

SVMs belong to a family of generalized linear models which achieves a classification or regression decision based on the value of the linear combination of features. They are also said to belong to "kernel methods". In addition to its solid mathematical foundation in statistical learning theory, SVMs have demonstrated highly competitive performance in numerous real-world applications, such as medical diagnosis, bioinformatics, face recognition, image processing and text mining, which has established SVMs as one of the most popular, state-of-the-art tools for knowledge discovery and data mining. Similar to artificial neural networks, SVMs possess the well-known ability of being universal approximators of any multivariate function to any desired degree of accuracy. Therefore, they are of particular interest to modeling highly nonlinear, complex systems and processes.

Generally, many linear classifiers (hyperplanes) are able to separate data into multiple classes. However, only one hyperplane achieves maximum separation. SVMs classify data as a part of a machine-learning process, which "learns" from the historic cases represented as data points. These data points may have more than two dimensions. Ultimately, we are interested in whether we can separate data by an *n*-1 dimensional hyperplane. This may be seen as a typical form of linear classifier. We are interested in finding if we can achieve maximum separation (margin) between the two (or more) classes. By this we mean that we pick the hyperplane so that the distance from the hyperplane to the nearest data point is maximized. Now, if such a hyperplane exists, the hyperplane is clearly of interest and is known as the maximum-margin hyperplane and such a linear classifier is known as a maximum margin classifier.

To run a support vector model in R on the training set:

```
install.packages('e1071', repos = "http://cran.us.r-proj
ect.org")

library(e1071)

training.index      =      sample(c(1:dim(CCreditRisk)[1]),
dim(CCreditRisk)[1]*0.8)
training.data = CCreditRisk[training.index,]
test.data = CCreditRisk[-training.index,]
```

```
classifier = svm(formula = loan_status ~ person_age +
person_income + loan_grade+ loan_amnt+ loan_percent_
income+ cb_person_default_on_file+cb_person_cred_hist_
length, data = training.data, type = 'C-classification',
kernel = 'linear')
```

To predict with this model in R:

```
sv_pred = predict(classifier, newdata = test.data[, -c(3,
4,5,8,9)])
```

This yields the coincidence matrix:

```
confusionMatrix(data=sv_pred, reference = test.data$loan_
status, positive = "1")

## Confusion Matrix and Statistics
##
##           Reference
## Prediction    0    1
##          0 4728  811
##          1  370  608
##
##                Accuracy : 0.8188
##                  95% CI : (0.8092, 0.8281)
##     No Information Rate : 0.7823
##     P-Value [Acc > NIR] : 1.644e-13
##
##                   Kappa : 0.4008
##
##  Mcnemar's Test P-Value : < 2.2e-16
##
##             Sensitivity : 0.42847
##             Specificity : 0.92742
##          Pos Pred Value : 0.62168
##          Neg Pred Value : 0.85358
##              Prevalence : 0.21774
##          Detection Rate : 0.09329
##    Detection Prevalence : 0.15007
##       Balanced Accuracy : 0.67795
```

```
##
##          'Positive' Class : 1
##
```

```
sv_prob = as.numeric(predict(classifier, test.data[, -c(3,4,
5,8,9)]))
par(pty="s")
test_roc = roc(test.data$loan_status ~ sv_prob, plot = TRUE,
print.auc = TRUE, legacy.axes=TRUE, percent=TRUE, xlab=
"False Positive Rate", ylab="True Positive Rate", lwd=4)

## Setting levels: control = 0, case = 1

## Setting direction: controls < cases

#####AUC##############
auc(test_roc)

## Area under the curve: 67.79%
```

The AUC plot for the SVM model is given in Figure 7.2.

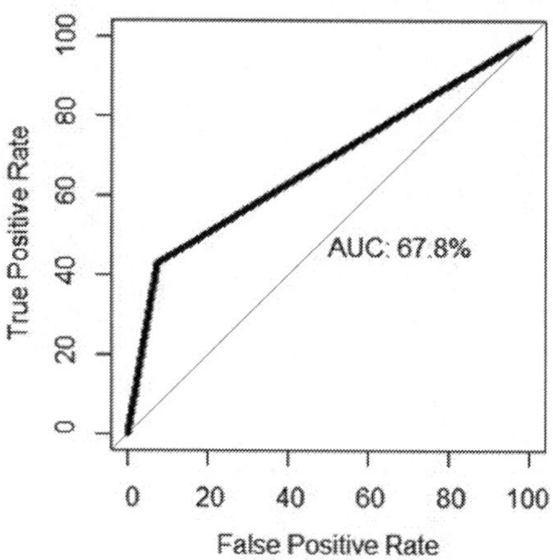

FIGURE 7.2
AUC for the SVM Model.

7.4 Neural Networks

Neural network models can also be applied to data that can be analyzed by alternative models for classification as well as regression. The normal data mining process is to try all alternative models and see which works best for a specific type of data over time. But there are some types of data where neural network models usually outperform alternatives, such as regression or decision trees. Neural networks tend to work better when there are complicated relationships in the data, such as high degrees of nonlinearity. They tend to be viable models in problem domains where there are high levels of unpredictability.

Each node is connected by an arc to nodes in the next layer. These arcs have weights, which are multiplied by the value of incoming nodes and summed. The input node values are determined by variable values in the dataset. Middle layer node values are the sum of incoming node values multiplied by the arc weights. These middle node values in turn are multiplied by the outgoing arc weights to successor nodes. Neural networks "learn" through feedback loops. For a given input, the output for starting weights is calculated. Output is compared to target values, and the difference between attained and target output is fed back to the system to adjust the weights on arcs.

This process is repeated until the network correctly classifies the proportion of learning data specified by the user (tolerance level). Ultimately, a set of weights might be encountered that explain the learning (training) dataset very well. The better the fit that is specified, the longer the neural network will take to train, although there is really no way to accurately predict how long a specific model will take to learn. The resulting set of weights from a model that satisfies the set tolerance level is retained within the system for application to future data.

The neural network model is a black box. To run a neural network model on the training data in R:

```
install.packages("nnet", repos = "http://cran.us.r-proj
ect.org")

library(nnet)
nn <- nnet(loan_status ~ person_age + person_income + loan_
grade+ loan_amnt+ loan_percent_income+ cb_person_default_
on_file+cb_person_cred_hist_length, data = training.data,
size=30, maxit=500, rang=0.1, decay=5e-4)

## # weights:  421
## initial  value 20006.671136
## iter  10 value 11971.175718
```

```
## iter  20 value 11764.599880
## iter  30 value 11622.064891
## iter  40 value 11617.332210
## iter  50 value 11616.582680
## iter  60 value 11614.601203
## iter  70 value 11613.997209
## iter  80 value 11613.827450
## final  value 11613.812725
## converged

install.packages("nnet", repos = "http://cran.us.r-proj
ect.org")

library(nnet)
nn <- nnet(loan_status ~ person_age + person_income + loan_
grade+ loan_amnt+ loan_percent_income+ cb_person_default_
on_file+cb_person_cred_hist_length, data = training.data,
size=30, maxit=500, rang=0.1, decay=5e-4)

## # weights:  421
## initial  value 20006.671136
## iter  10 value 11971.175718
## iter  20 value 11764.599880
## iter  30 value 11622.064891
## iter  40 value 11617.332210
## iter  50 value 11616.582680
## iter  60 value 11614.601203
## iter  70 value 11613.997209
## iter  80 value 11613.827450
## final  value 11613.812725
## converged
```

To predict with this model:

```
ann_pred <- as.factor(predict(nn, test.data, type=
"class"))
```

This yields the confusion matrix on the test set:

```
confusionMatrix(data=ann_pred, reference = test.data$loan_
status, positive = "1")

## Confusion Matrix and Statistics
##
##           Reference
## Prediction    0    1
##          0 4862  904
##          1  236  515
##
##                Accuracy : 0.8251
##                  95% CI : (0.8156, 0.8342)
##     No Information Rate : 0.7823
##     P-Value [Acc > NIR] : < 2.2e-16
##
##                   Kappa : 0.3814
##
##  Mcnemar's Test P-Value : < 2.2e-16
##
##             Sensitivity : 0.36293
##             Specificity : 0.95371
##          Pos Pred Value : 0.68575
##          Neg Pred Value : 0.84322
##              Prevalence : 0.21774
##          Detection Rate : 0.07902
##    Detection Prevalence : 0.11524
##       Balanced Accuracy : 0.65832
##
##        'Positive' Class : 1
##
```

To predict with this model:

```
test_prob = (predict(nn, test.data))
par(pty="s")
test_roc = roc(test.data$loan_status ~ test_prob,
plot = TRUE, print.auc = TRUE, legacy.axes=TRUE, percent=
TRUE, xlab="False Positive Rate", ylab="True Positive
Rate", lwd=4)
```

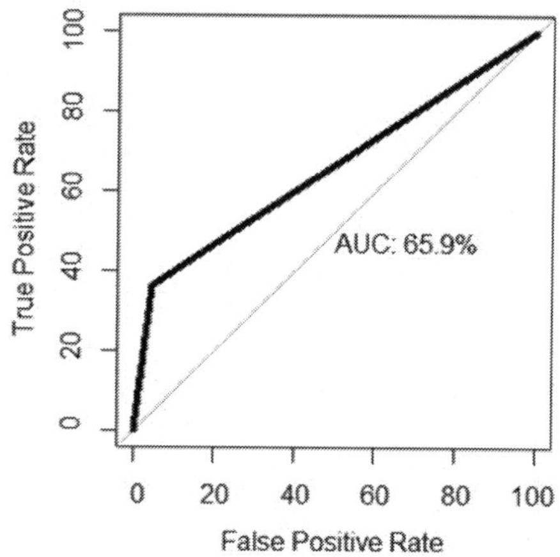

FIGURE 7.3
AUC for the Neural Network Model.

To obtain the area under the curve (Figure 7.3).

```
#####AUC#############
auc(test_roc)
## Area under the curve: 65.86%
```

7.5 Decision Trees

Decision trees are also called classification and regression trees (CART) and can provide a means to obtain product-specific prediction models in the form of rules that are easy to implement. These rules have an IF-THEN form, which is easy for users to implement. This data mining approach can be used by groceries in a number of policy decisions, to include ordering inventory replenishment as well as evaluation of alternative promotion campaigns.

As was the case with regression models and neural networks, decision tree models support the data mining process of modeling. Decision trees in the context of data mining refer to the tree structure of rules. The data mining decision tree process involves collecting those variables that the analyst thinks might bear on the decision at issue and analyzing these variables for their ability to predict outcome. Decision trees are useful to gain further insight into customer behavior as well as lead to ways to profitably act on results. The algorithm automatically determines which variables are most important, based on their ability to sort the data into the correct output category. The method has relative advantage over neural network and genetic algorithms in that a reusable set of rules are provided, thus explaining model conclusions. There are many examples where decision trees have been applied to business data mining, including classifying loan applicants, screening potential consumers, and rating job applicants.

Decision trees provide a way to implement rule-based system approaches. All objects in the training set are classified into these branches. If all objects in a branch belong to the same output class, the node is labeled, and this branch is terminated. If there are multiple classes on a branch, another attribute is selected as a node, with all possible attribute values branched. An entropy heuristic is used to select the attributes with the highest information. In other data mining tools, other bases for selecting branches are often used.

To run a decision tree model in R with the loan dataset y:

```
dt <- rpart(loan_status ~ person_age + person_income + loan_
grade+ loan_amnt+ loan_percent_income+ cb_person_default_
on_file+cb_person_cred_hist_length, data = training.data,
method="class")

#plot(as.party(dt))
prp(dt, type = 4, extra = 6, box.palette = "auto")
```

Figure 7.4 gives a schematic of the R decision tree model.

The fit parameters are obtained as follows:

```
# print the complexity parameters and plot the error vs
complexity
printcp(dt)

##
## Classification tree:
## rpart(formula = loan_status ~ person_age + person_
income + loan_grade +
##    loan_amnt + loan_percent_income + cb_person_default_
on_file +
```

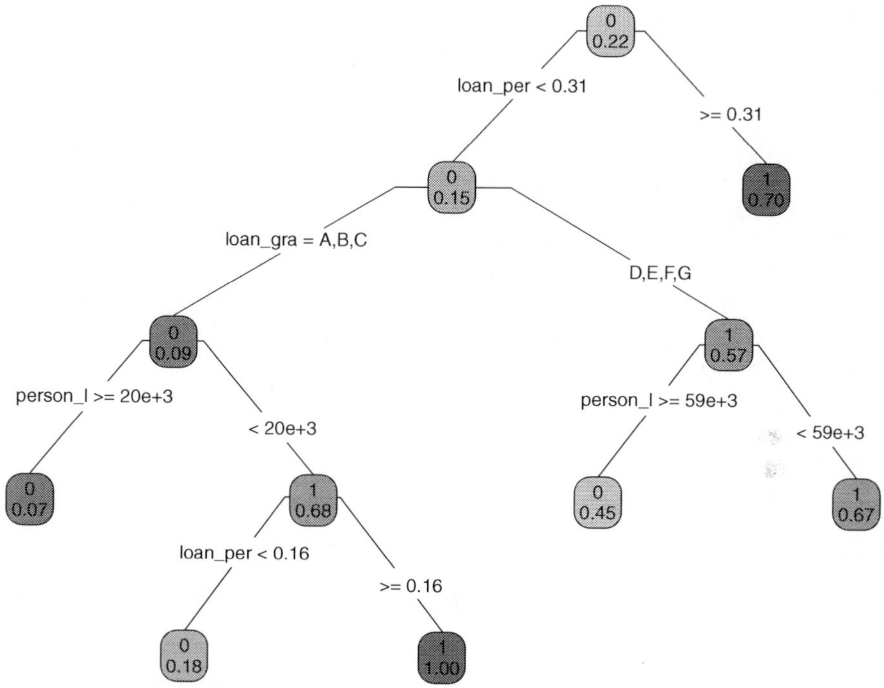

FIGURE 7.4
Decision Tree from R.

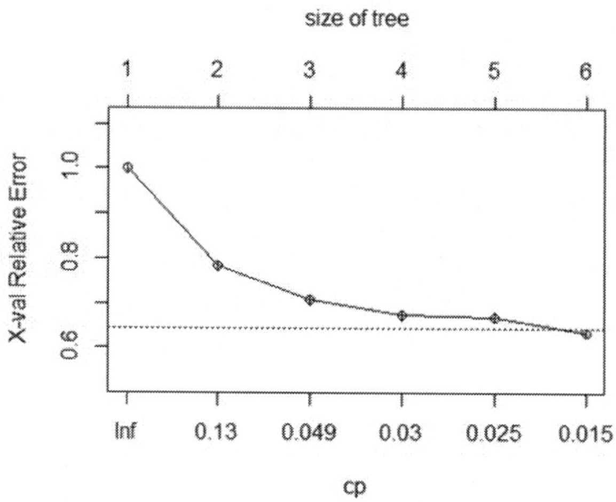

FIGURE 7.5
Tree Size versus Relative Error.

```
##      cb_person_cred_hist_length, data = training.data,
method = "class")
##
## Variables actually used in tree construction:
## [1] loan_grade              loan_percent_income person_
income
##
## Root node error: 5666/26064 = 0.21739
##
## n= 26064
##
##         CP nsplit rel error  xerror      xstd
## 1 0.216555      0   1.00000 1.00000 0.0117526
## 2 0.076421      1   0.78345 0.78345 0.0107108
## 3 0.031945      2   0.70702 0.70702 0.0102764
## 4 0.028062      3   0.67508 0.67367 0.0100739
## 5 0.021532      4   0.64702 0.66996 0.0100509
## 6 0.010000      5   0.62549 0.63519 0.0098299
```

The relationship of tree size with relative error is displayed in Figure 7.5.

```
plotcp(dt)
```

Error fits are obtained as follows:

```
DT_pred <- predict(dt, test.data, decision.values=TRUE,
type="class")
confusionMatrix(data=DT_pred, reference = test.data$loan_
status, mode = "everything", positive = "1")

## Confusion Matrix and Statistics
##
##           Reference
## Prediction    0    1
##          0 4681  513
##          1  394  929
##
##                Accuracy : 0.8608
##                  95% CI : (0.8522, 0.8691)
```

```
##       No Information Rate : 0.7787
##       P-Value [Acc > NIR] : < 2.2e-16
##
##                     Kappa : 0.5839
##
##    Mcnemar's Test P-Value : 8.924e-05
##
##               Sensitivity : 0.6442
##               Specificity : 0.9224
##            Pos Pred Value : 0.7022
##            Neg Pred Value : 0.9012
##                 Precision : 0.7022
##                    Recall : 0.6442
##                        F1 : 0.6720
##                Prevalence : 0.2213
##            Detection Rate : 0.1426
##      Detection Prevalence : 0.2030
##         Balanced Accuracy : 0.7833
##
##          'Positive' Class : 1
##
```

To obtain the area under the curve (Figure 7.6).

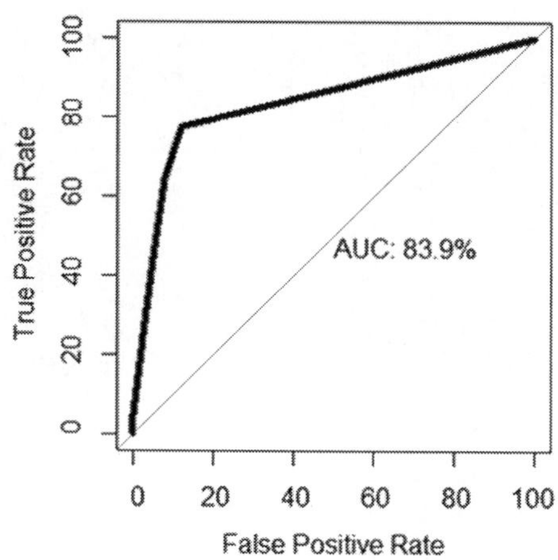

FIGURE 7.6
AUC for Decision Tree Model.

```
par(pty="s")
test_prob = predict(dt, test.data, type = "prob")[,2]
test_roc    =    roc(test.data$loan_status    ~    test_prob,
plot = TRUE, print.auc = TRUE, legacy.axes=TRUE, percent=
TRUE,  xlab="False  Positive  Rate",  ylab="True  Positive
Rate", lwd=4)

## Setting levels: control = 0, case = 1
## Setting direction: controls < cases

#####AUC#############
auc(test_roc)

## Area under the curve: 83.92%
```

7.6 Random Forests

Random forests are ensembles collecting un-pruned decision trees (typically tens or hundreds of decision trees), providing models that tend to be more robust with respect to variance and bias. Random forests are attractive when there are large training datasets and when there are many input variables. The random forest model from R is obtained with:

```
## Random Forest
install.packages("randomForest",  repos  =  "http://cran.
us.r-project.org")

library(randomForest)

rf=randomForest(loan_status   ~   person_age   +   person_
income  +  loan_grade+  loan_amnt+  loan_percent_income+
cb_person_default_on_file+cb_person_cred_hist_length,
data = training.data, ntree=500, nodesize= 5, importance=
TRUE)
varImpPlot(rf, type=1)
```

Figure 7.7 displays accuracy contributions by variable.

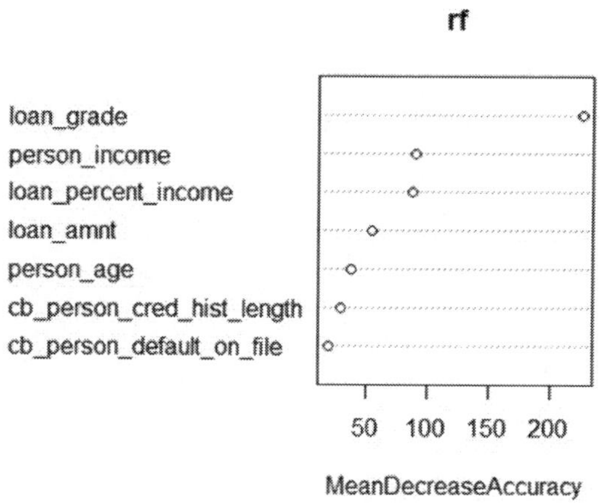

FIGURE 7.7
Random Forest Accuracy Progress.

The relative fit measures for the random forest model are obtained as follows:

```
rf.pred=predict(rf, test.data)
confusionMatrix(data=rf.pred, reference = test.data$loan_
status, positive = "1")

## Confusion Matrix and Statistics
##
##           Reference
## Prediction    0    1
##          0 4763  519
##          1  312  923
##
##                Accuracy : 0.8725
##                  95% CI : (0.8641, 0.8805)
##     No Information Rate : 0.7787
##     P-Value [Acc > NIR] : < 2.2e-16
##
##                   Kappa : 0.6099
##
##  Mcnemar's Test P-Value : 8.93e-13
##
##             Sensitivity : 0.6401
##             Specificity : 0.9385
##          Pos Pred Value : 0.7474
```

```
##          Neg Pred Value : 0.9017
##              Prevalence : 0.2213
##          Detection Rate : 0.1416
##    Detection Prevalence : 0.1895
##       Balanced Accuracy : 0.7893
##
##        'Positive' Class : 1
##
```

7.7 Boosting

Boosting is a means to run multiple models with different weights to reduce error, much in the spirit of neural networks, or as we have just seen, random forests. All involve intense computation searching for reduced error over the training set. Boosting associates a weight to each of the datasets observations, building a series of models with weights increased if it incorrectly classifies that observation. This yields a series of decision trees in the ensemble. Extreme boosting uses the extreme gradient boosting algorithm which provides better fits in most cases.

To run an extreme boosting model in R:

```
install.packages("xgboost", repos = "http://cran.us.r-
project.org")

library(xgboost)
library(Matrix)
library(magrittr)

CCreditRisk = as.data.frame(read.csv("Chinacredit_risk_
dataset.csv"))
#variable conversion-
CCreditRisk$loan_status=as.factor(CCreditRisk$loan_
status)
CCreditRisk$person_home_ownership = as.
factor(CCreditRisk$person_home_ownership)
CCreditRisk$loan_intent = as.factor(CCreditRisk$loan_
intent)
CCreditRisk$loan_grade = as.factor(CCreditRisk$loan_
grade)
```

```
CCreditRisk$cb_person_default_on_file =as.
factor(CCreditRisk$cb_person_default_on_file)

training.index = sample(c(1:dim(CCreditRisk)[1]),
dim(CCreditRisk)[1]*0.8)

training.data = CCreditRisk[training.index,]
test.data = CCreditRisk[-training.index,]
training.data = training.data[, -c(3,4,5,8)]
test.data=test.data[, -c(3,4,5,8)]

sparse_matrix <- sparse.model.matrix(training.data$loan_
status ~ ., data = training.data)[,-1]

sparse_matrix_test <- sparse.model.matrix(test.data$loan_
status ~ ., data = test.data)[,-1]
dtrain <- xgb.DMatrix(sparse_matrix, label = as.
integer(training.data$loan_status)-1)
dtest<- xgb.DMatrix(sparse_matrix_test, label = as.
integer(test.data$loan_status)-1)
xgmodel <- xgboost(data = dtrain, max.depth = 2, eta = 1,
nthread = 2, nrounds = 2, objective = "binary:logistic")
## [1]  train-logloss:0.415180
## [2]  train-logloss:0.365751
```

The relative fit of the extreme boosting model is obtained as follows:

```
Xgpred=predict(xgmodel, dtest)
Xgpred=as.factor(ifelse(Xgpred>0.5, 1, 0))
confusionMatrix(data=Xgpred , reference = test.data$loan_
status, positive = "1")

## Confusion Matrix and Statistics
##
##           Reference
## Prediction    0    1
##          0 4780  684
##          1  282  771
##
##                Accuracy : 0.8518
```

```
##                95% CI : (0.8429, 0.8603)
##     No Information Rate : 0.7767
##     P-Value [Acc > NIR] : < 2.2e-16
##
##                   Kappa : 0.526
##
##  Mcnemar's Test P-Value : < 2.2e-16
##
##             Sensitivity : 0.5299
##             Specificity : 0.9443
##          Pos Pred Value : 0.7322
##          Neg Pred Value : 0.8748
##              Prevalence : 0.2233
##          Detection Rate : 0.1183
##    Detection Prevalence : 0.1616
##       Balanced Accuracy : 0.7371
##
##        'Positive' Class : 1
```

7.8 Comparison

Running all of these models gives a flavor for the R toolbox of classification models. Table 7.3 compares test results for these models using 80 percent of the data for training and holding 20 percent out for testing.

Specificity (Spec) is the ratio of correct predictions when the model predicts no default, while sensitivity (Sens) is the ratio of correct predictions when

TABLE 7.3

Comparison of Coincidence Matrices

Model	Actual no Model no	Actual no Model yes	Spec	Actual yes Model no	Actual yes Model yes	Sens	Overall
Log Reg	4789	286	0.944	749	693	0.481	0.841
SVM	4728	370	0.927	811	608	0.428	0.819
Neural net	4862	236	0.954	904	515	0.363	0.825
D Tree	4681	394	0.922	513	929	0.644	0.861
Random For	4763	312	0.939	519	923	0.640	0.872
Boosting	4780	282	0.944	684	771	0.530	0.852

TABLE 7.4

Comparison – Errors Weighted by Cost

Model	Actual no Model yes Cost 1	Actual yes Model no Cost 5	Total Cost
Log Reg	286	749	4031
SVM	370	811	4425
Neural Net	236	904	4756
Decision Tree	394	513	2959
Random For	282	519	2877
Boosting	312	684	3733

the model predicts default. Overall is the ratio of correct predictions for both classes of prediction. By specificity, all models were quite accurate, with the neural network best at predicting OK (not default) outcomes. The best model at predicting default test cases by sensitivity was the decision tree. Sensitivity measures the test case prediction rate for defaults. Overall prediction was strongest for random forest models.

Usually, relative cost matters a lot. If each case that the model predicted to be bankrupt actually wasn't, business would have been lost – the expected profit of that investment. The other type of error would be to invest in a firm that actually went bankrupt. If the cost of this case was 5 times that of the first type of error, we could compare relative costs (Table 7.4).

Considering weighted costs, the random forest and decision tree models were best. Note that it is usually best practice to run multiple models and see which fit your particular dataset best, realizing that the results can change with new data.

7.9 Conclusions

We recap the essential points in classification modeling:

1. The process of training, testing, and applying the model is best done if you have independent training and test sets. In this case, you do not want to partition the data. If you need to skip the test set, you can partition, selecting the proportion of the dataset you want the software to hold out for testing.
2. Classification is for categorical target variables. Prior chapters have covered techniques for continuous target variables. Ordinary least squares is the common regression model for continuous variables. Logistic regression works for categorical targets. Decision trees work

best for categorical targets. You can also run regression trees on continuous targets, but as we discussed in that chapter, the finite number of possible predictions make regression trees less believable.

3. Models can be controlled. Regression is controlled primarily through variable selection. Using all available input variables will always yield a higher R^2, but will usually involve overlapping information that makes the β coefficients unstable. Predicting isn't a problem as long as you have inputs for all independent variables, but trimmed models are stabler. Decision trees are controlled by setting minimum support and minimum confidence. An easier control can be done by using the complexity parameter. A lower complexity will yield models with fewer rules, which are easier to implement. Neural network models can be controlled by adjusting hidden layers, although "control" is probably too clear a term in this case.
4. There are alternative softwares available. Commercial software is very good, but often very expensive. Rattle and WEKA are open source. There are quite a few other open source softwares. WEKA is very interesting from a research perspective, with many optional modeling techniques. Rattle is a GUI for R, which makes it commercially viable and scalable.

8

Fraud Detection

Fraud detection is one of the most widely applied applications of data mining. It is usually a classification problem, with outcomes of fraud or not, which hopefully is highly biased to not. Fraud detection has been applied in many fields, one of the most common being credit card fraud detection (Alwan et al., 2022). But it also has been applied in customs fraud (Vanhoeyveld et al., 2020), insider trading detection (Fevzi Esen et al., 2019, fraud in public procurement (Velasco et al., 2021), and detection of stock market manipulation (Siering et al., 2021). A common feature of fraud detection is dataset imbalance. This chapter will apply the random over-sampling examples (ROSE) method in R to balance credit card fraud data.

8.1 Types of Fraud

8.1.1 Credit Card Fraud Dataset

Credit cards (in a number of forms, to include bank cards) are replacing cash. This provides a great deal of convenience and efficiency, but also provides fraudsters with an arena to work in. Credit card fraud can be accomplished offline by robbery of cards, or most often on-line over the Internet or cell phone system. Alwan et al. (2022) focused on four types of credit card fraud:

1. Postcards – credit is widely offered through the mail. Postcards offering credit card accounts are often taken from mailboxes or garbage bins and submitted in a form of identity theft.
2. Replication from websites – it is very convenient and efficient to make purchases over the Internet, and hackers can pirate card accounts and passwords in another form of identity theft.
3. Phishing – e-mail and telephone phishing is an industry that makes cell phone use irritating, and also leads to capture of accounts and passwords from careless or over-trusting legitimate users.

DOI: 10.1201/9781003562399-8

4. Triangulation – fraudsters can create websites and ads offering very attractive purchases over the Internet, using credit card information. The website operators can then use this information to make illegitimate purchases.

Challenges in detecting credit card fraud include dealing with unbalanced data (in that to date at least, fraudulent cases are rare relative to legitimate purchases, warping the statistical database). The relative cost of error is a factor, as detecting a fraudulent case saves a small amount without generating revenue, at least until the moral hazard factor takes effect. Another difficulty is that fraudsters continue to innovate and develop new approaches. The most common tools to detect credit card fraud are classification models.

8.1.2 Customs Fraud

International trade has boomed as global supply chains have revolutionized modern business. Thus, there is a massive increase in the volume, variety, and complexity of goods crossing national borders. Customs are responsible for checking the legitimacy of international transactions and collection of appropriate fees.

Types of customs fraud include product misclassification (to pay tariffs at lower rates), fraudulent valuation, smuggling (to avoid tariffs altogether), and manipulation of goods origin (to obtain more favorable tariff rates). There is also a great deal of illegal drug trafficking and importing of counterfeit goods.

Vanhoeyveld et al. (2020) analyzed Belgian customs data. Customs fraud was found to occur for goods by particular entities. Classification algorithms were applied. A typical problem of fraud detection data is the presence of skewness. A support vector machine model was found to provide a useful tool to identify customs fraud at a data-level involving consignee, declarant, commodity, and country of origin.

8.1.3 Corporate Insider Trading

Financial fraud includes insider trading, where corporate executives and employees have access to information enabling them to make higher returns than the public with informational disadvantage. Insider trading is defined as buying or selling securities by insiders while in possession of material, non-public information which is not available to the public. Captains of industry in the 19th Century became fabulously wealthy, often if not always through effective utilization of insider information. The legal system has ever since worked hard to stop unfair advantage, but a great deal of intelligence is applied on the other side to come up with new ways to take advantage of the investing public.

Detection of insider trading again is often analyzed with classification models (Fevzi Esen et al., 2019). The target is usually the identification of abnormal returns. There is also a great deal of effort applied with text mining, analysis of documents such as e-mails or instructions to traders.

A special form of stock market fraud is market manipulation, where fraudsters publish highly positive false information through websites, spam messages, and other forms of advertising campaigns (Siering et al., 2021). They benefit through strategies of buying low and spamming high. Kasztelnik and Jernakowicz (2024) reviewed notable recent cases in Germany, China, and Japan. Text mining is a useful tool to identify stock manipulation, followed by the application of classification algorithms.

Kasztelnik and Jernakowicz (2024) reported on the potential of artificial intelligence to detect financial statement fraud. While machine learning models have strong potential to automatically detect fraudulent patterns, they also involve some risks:

- The risk of overreliance on AI technology may create a false sense of security. There is a need to monitor output and apply professional judgment, experience, and skepticism.
- False positives are encountered, flagging legitimate financial activity as fraudulent leading to unnecessary investigation, cost, and damage to reputations.
- Limited training data can be a problem in initial applications of AI and data analytics. Their effectiveness depends on the quality of the data used.
- Cybersecurity – use of AI can expose financial data to hacking and data breaches.
- Ethics – there is potential for privacy and bias that can lead to unintended consequences.

The field of financial security fraud is important because of the amounts of potential profit available to fraudsters. There is an on-going game between fraudsters and regulators that will continue as long as there is human trading.

8.1.4 Public Procurement Fraud

Governments purchase enormous amounts of goods and services. The volume of money involved attracts some fraudulent activity. Corruption and money laundering occur throughout the world. Velasco et al. (2021) researched public procurement fraud in Brazil, applying classification models for its detection. A variety of databases were combined for the study.

Collusion patterns were detected to include identical bids, indicating possible prearrangement, or with identical ratios or differences. Sometimes a shell company will submit a bid that is just high enough to lose, thus providing support to the validity of the lowest bidder. Common partners, addresses belonging to the same parent company, and with the same IP address are also found on occasion. Velasco et al. (2021) gathered information on bidding company black listings, links to political campaigns, and the presence of high-cost overruns were also indicators of fraud. Conflicts of interest from relatives of public officials was also found to be a negative indicator. Identification of fraud indicators was accomplished through the use of graph theoretical models and network analysis.

8.1.5 Fraud in Social Networks

The rapid development of global communication technology has been accompanied by fraud. Yang et al. (2021) found phone fraud in China to be significant. Fraudsters often play the role of legal officials and swindle gullible victims.

Yang et al. (2021) also examined ego-network characteristics (ratio of outgoing to incoming calls), finding fraudsters, although they also have a high number of incoming calls. Energy dispersion is the proportion of caller activity with a particular contact. Fraudsters had a large set of connections (potential targets). The fraction of fraudster reciprocal calls was found to be lower than that of non-fraudsters. Fraudsters did not make random calls but focused on user age and mobile activity.

The study used demographic, ego-network, and call behavior data. Models used for detection were logistic regression, support vector machines, and graph-based models. Balancing was accomplished with the random oversampling examples (ROSE) method, an alternative to synthetic minority over-sampling technique (SMOTE). Both draw observations from the minority class to provide better dataset balancing, which improves statistical performance.

8.2 Credit Card Fraud Dataset

Digital payments are evolving, but so are cyber criminals. According to the Data Breach Index, more than 5 million records are being stolen daily, a concerning statistic that shows fraud is still very common both for Card-Present and Card-Not-Present type of payments. The following data (n=1,000,000) was obtained from Kaggle.com (available from the authors) and is described in Table 8.1.

Of the million observations, 87,403 are fraudulent.

TABLE 8.1

Credit Card Fraud Variables

Variable	Description	Type	Min	Max
Distance_from_home	home to transaction point	number	0.0048	10632
Distance_from_last_transaction	To last transaction	number	0.0001	11851
Ratio_to_median_purchase_price	Ratio	number	0.0044	267.8
Repeat_retailer		Binary		
Used_chip	Credit card chip used	Binary		
Used_pin_number		Binary		
Online_order		Binary		
Fraud	Transaction is fraudulent	Binary		

8.2.1 R Analysis

The first step is to link the data file:

```
##READ THE DATA FILE ###
dataCredit <- as.data.frame(read.csv("CreditCardFraud.csv"))
#check variable types and make variable conversion
str(dataCredit)

## 'data.frame':    1000000 obs. of  8 variables:
##  $ DistanceHome: num  57.88 10.83 5.09 2.25 44.19 ...
##  $ DistanceLast: num  0.311 0.176 0.805 5.6 0.566 ...
##  $ RatioMedian : num  1.946 1.294 0.428 0.363 2.223 ...
##  $ Repeat      : int  1 1 1 1 1 1 1 1 0 1 ...
##  $ UsedChip    : int  1 0 0 1 1 0 0 0 0 0 ...
##  $ UsedPin     : int  0 0 0 0 0 0 0 1 0 0 ...
##  $ Online      : int  0 0 1 1 1 0 1 0 0 1 ...
##  $ Fraud       : int  0 0 0 0 0 0 0 0 0 0 ...

summary(dataCredit)
##   DistanceHome         DistanceLast         RatioMedian           Repeat
## Min.   :    0.005   Min.   :    0.000   Min.   :  0.0044   Min.   :0.0000
## 1st Qu.:    3.878   1st Qu.:    0.297   1st Qu.:  0.4757   1st Qu.:1.0000
## Median :    9.968   Median :    0.999   Median :  0.9977   Median :1.0000
## Mean   :   26.629   Mean   :    5.037   Mean   :  1.8242   Mean   :0.8815
## 3rd Qu.:   25.744   3rd Qu.:    3.356   3rd Qu.:  2.0964   3rd Qu.:1.0000
## Max.   :10632.724   Max.   :11851.105   Max.   :267.8029   Max.   :1.0000
##     UsedChip           UsedPin             Online             Fraud
## Min.   :0.0000     Min.   :0.0000     Min.   :0.0000     Min.   :0.0000
## 1st Qu.:0.0000     1st Qu.:0.0000     1st Qu.:0.0000     1st Qu.:0.0000
## Median :0.0000     Median :0.0000     Median :1.0000     Median :0.0000
## Mean   :0.3504     Mean   :0.1006     Mean   :0.6506     Mean   :0.0874
## 3rd Qu.:1.0000     3rd Qu.:0.0000     3rd Qu.:1.0000     3rd Qu.:0.0000
## Max.   :1.0000     Max.   :1.0000     Max.   :1.0000     Max.   :1.0000
```

```
#
#dataCredit$Repeat=as.factor(dataCredit$Repeat)
#dataCredit$UsedChip= as.factor(dataCredit$UsedChip)
#dataCredit$UsedPin= as.factor(dataCredit$UsedPin)
#dataCredit$OnLine= as.factor(dataCredit$OnLine)
dataCredit$Fraud= as.factor(dataCredit$Fraud)
sum(dataCredit$Fraud == 1)

## [1] 87403
```

The dataset of 1,000,000 observations had 87,403 fraudulent cases, which is highly imbalanced. We use ROSE package in R to execute data balancing operations. In the ROSE package, over-sampling, under-sampling and both are options for data balancing. In R this is done as follows:

```
install.packages("ROSE", repos = "http://cran.us.r-proj
ect.org")

## Installing package into 'C:/Users/oaraz2/AppData/
Local/R/win-library/4.4'
## (as 'lib' is unspecified)

## package 'ROSE' successfully unpacked and MD5 sums checked
##
## The downloaded binary packages are in
##  C:\Users\oaraz2\AppData\Local\Temp\RtmpgJROaz\
downloaded_packages

library(ROSE)

## Loaded ROSE 0.0-4

#Data balancing

#1 Over-sampling for data balancing p=0.5
set.seed(123)
balanced_data=ovun.sample(Fraud~., data = dataCredit,
method = "over", p=0.5)$data
table(balanced_data$Fraud)

##
##      0      1
## 912597 912165
```

ROSE generated an expanded dataset with 912,165 fraudulent cases. This number will vary depending upon the seed that is set. The goal is to have roughly the same number of OK and fraud cases.

This dataset is now partitioned to have 80 percent training and 20 percent held out for testing.

```
train.index     <-     sample(c(1:dim(balanced_data)[1]),
dim(balanced_data)[1]*0.8)
train.df <- balanced_data[train.index, ]
valid.df <- balanced_data[-train.index, ]
```

The next step is to run classification models. We begin with logistic regression:

8.2.2 Logistic Regression Model

```
# run logistic regression
# use glm() (general linear model) with family = "bino-
mial" to fit a logistic
# regression.
B_fraud_model <- glm(Fraud ~ ., data = train.df,
family = "binomial")

## Warning: glm.fit: fitted probabilities numerically 0 or
1 occurred
```

R yielded the logistic regression coefficients shown below. All were highly significant.

```
options(scipen=999)
summary(B_fraud_model)

##
## Call:
## glm(formula = Fraud ~ ., family = "binomial", data = train.df)
##
## Coefficients:
##                  Estimate  Std. Error z value               Pr(>|z|)
## (Intercept)   -7.62978359  0.01851707  -412.0 <0.0000000000000002 ***
## DistanceHome   0.02890942  0.00007677   376.6 <0.0000000000000002 ***
## DistanceLast   0.04958375  0.00019549   253.6 <0.0000000000000002 ***
```

```
## RatioMedian     1.20868738  0.00230440   524.5 <0.0000000000000002 ***
## Repeat         -1.43429942  0.01084834  -132.2 <0.0000000000000002 ***
## UsedChip       -1.19467321  0.00780045  -153.2 <0.0000000000000002 ***
## UsedPin       -10.07682259  0.05540033  -181.9 <0.0000000000000002 ***
## Online          5.01287782  0.01478223   339.1 <0.0000000000000002 ***
## ---
## Signif. codes:  0 '***' 0.001 '**' 0.01 '*' 0.05 '.' 0.1 ' ' 1
##
## (Dispersion parameter for binomial family taken to be 1)
##
##     Null deviance: 2023724  on 1459808  degrees of freedom
## Residual deviance:  608573  on 1459801  degrees of freedom
## AIC: 608589
##
## Number of Fisher Scoring iterations: 10
```

Predictions with this model can be obtained with the following syntax. The first 10 datapoints are displayed here (observations 18, 52, and 56 were predicted to be fraudulent but weren't).

```
# use predict() with type = "response" to compute
predicted probabilities.
fraud.pred <- predict(B_fraud_model, valid.df[, -8],
type = "response")

# first 10 actual and predicted records
data.frame(actual = valid.df$Fraud[1:10],
predicted = ifelse(fraud.pred[1:10]>0.5, 1,0))

##    actual predicted
## 5       0         0
## 7       0         0
## 14      0         0
## 15      0         0
## 16      0         0
## 18      0         1
## 22      0         0
## 39      0         0
## 52      0         1
## 56      0         1
```

Odds ratios are obtained as follows:

```
####ODDS RATIOS###
round(exp(coef(B_fraud_model)),3)

## (Intercept) DistanceHome DistanceLast  RatioMedian
Repeat     UsedChip
##       0.000        1.029        1.051        3.349
0.238        0.303
##     UsedPin       Online
##       0.000      150.337

## odds ratios and 95% CI
#exp(cbind(OR = coef(fraud_model), confint(fraud_model)))
###INTERPRETATION###

# Form a table from the output-- CONFUSION MATRIX ----

#install caret and use confusionMatrix()

### PLOTING AUCs#####
install.packages(c("rattle", "rpart", "rpart.plot",
"partykit", "caret", "pROC"), repos = "http://cran.us.r-
project.org")

## Installing packages into 'C:/Users/oaraz2/AppData/
Local/R/win-library/4.4'
## (as 'lib' is unspecified)

##
##   There is a binary version available but the source
version is later:
##       binary source needs_compilation
## caret 6.0-94  7.0-1              TRUE
##
##   Binaries will be installed

## package 'caret' successfully unpacked and MD5 sums
checked

## package 'pROC' successfully unpacked and MD5 sums
checked

# call packages and load
```

```
library(caret)

## Loading required package: ggplot2

## Loading required package: lattice

fraud.pred= factor(ifelse (fraud.pred >0.5, 1, 0))
confusionMatrix(data=fraud.pred, reference = valid.
df$Fraud, mode = "everything", positive = "1")

## Confusion Matrix and Statistics
##
##           Reference
## Prediction      0      1
##          0 171028   9210
##          1  12137 172578
##
##                Accuracy : 0.9415
##                  95% CI : (0.9407, 0.9423)
##     No Information Rate : 0.5019
##     P-Value [Acc > NIR] : < 0.0000000000000002
##
##                   Kappa : 0.883
##
##  Mcnemar's Test P-Value : < 0.0000000000000002
##
##             Sensitivity : 0.9493
##             Specificity : 0.9337
##          Pos Pred Value : 0.9343
##          Neg Pred Value : 0.9489
##               Precision : 0.9343
##                  Recall : 0.9493
##                      F1 : 0.9418
##              Prevalence : 0.4981
##          Detection Rate : 0.4729
##    Detection Prevalence : 0.5061
##       Balanced Accuracy : 0.9415
##
##        'Positive' Class : 1
##
```

R yielded the correct identification of the 181,788 fraudulent cases 172,578 times, for a recall of 0.9493. It falsely predicted fraud 12,137 times.

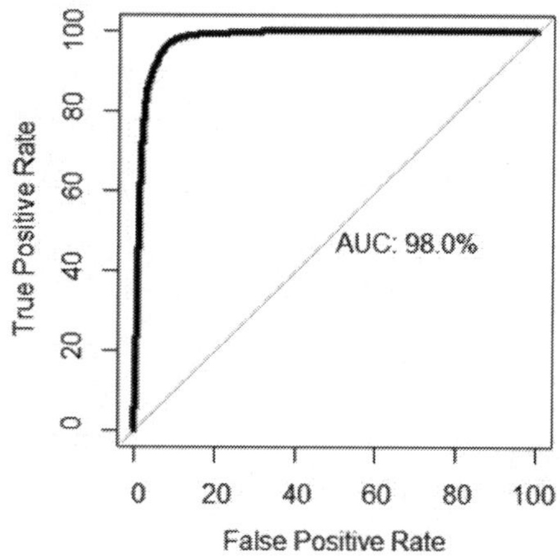

FIGURE 8.1
Area under the Curve for the Logistic Regression Model.

The area under the curve for the logistic regression model is obtained as follows yielding the plot in Figure 8.1:

```
#test_roc = roc(valid.df$Fraud ~ fraud.pred, plot = TRUE,
print.auc = TRUE)
###Fix the Plot area to Square####
par(pty="s")
test_roc = roc(valid.df$Fraud ~ fraud.pred, plot = TRUE,
print.auc = TRUE, legacy.axes=TRUE, percent=TRUE, xlab=
"False Positive Rate", ylab="True Positive Rate", lwd=4)

## Setting levels: control = 0, case = 1

## Setting direction: controls < cases
```

8.2.3 Decision Tree Model

The next model was a decision tree.

```
B_DT_fraud = rpart(Fraud ~., method="class", data=train.
df)
```

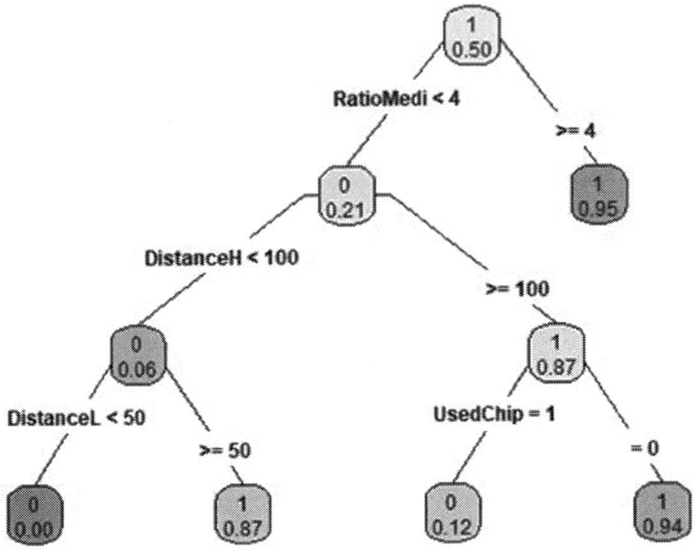

FIGURE 8.2
R Decision Tree.

The decision tree model obtained is shown in Figure 8.2:

```
#plot(as.party(DT_fraud))
prp(B_DT_fraud, type = 4, extra = 6, box.palette = "auto")
```

```
#fancyRpartPlot(DT_fraud)
# print the complexity parameters and plot the error vs
complexity
printcp(B_DT_fraud)

##
## Classification tree:
## rpart(formula = Fraud ~ ., data = train.df,
method = "class")
##
## Variables actually used in tree construction:
## [1] DistanceHome DistanceLast RatioMedian UsedChip
##
## Root node error: 729432/1459809 = 0.49968
```

```
##
## n= 1459809
##
##          CP nsplit rel error   xerror       xstd
## 1 0.701059      0  1.000000 1.000000 0.00082820
## 2 0.169107      1  0.298941 0.298942 0.00059043
## 3 0.049580      2  0.129834 0.129837 0.00040798
## 4 0.014513      3  0.080254 0.080258 0.00032499
## 5 0.010000      4  0.065742 0.066061 0.00029593

plotcp(B_DT_fraud)
```

Decrease in relative error by tree branch size is displayed in Figure 8.3:
The five rules obtained are printed using the following syntax:

```
#Extract the decision tree rules as text
rpart.rules(B_DT_fraud)
##  Fraud
##     0.00 when RatioMedian <  4 & DistanceHome <  100 &
DistanceLast <  50
##     0.12 when RatioMedian <   4 & DistanceHome >= 100
& UsedChip is 1
##     0.87 when RatioMedian <  4 & DistanceHome <  100 &
DistanceLast >= 50
```

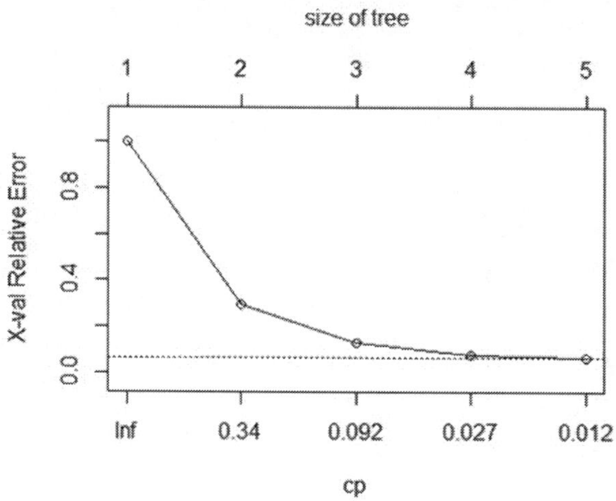

FIGURE 8.3
R Decision Tree Relative Error by Tree Size.

```
##    0.94 when RatioMedian <   4 & DistanceHome >= 100
& UsedChip is 0
##   0.95 when RatioMedian >= 4

asRules(B_DT_fraud)

##
##  Rule number: 3 [Fraud=1 cover=573544 (39%) prob=0.95]
##    RatioMedian>=4
##
##  Rule number: 11 [Fraud=1 cover=153030 (10%) prob=0.94]
##    RatioMedian< 4
##    DistanceHome>=100
##    UsedChip< 0.5
##
##  Rule number: 9 [Fraud=1 cover=48345 (3%) prob=0.87]
##    RatioMedian< 4
##    DistanceHome< 100
##    DistanceLast>=50
##
##  Rule number: 10 [Fraud=0 cover=13998 (1%) prob=0.12]
##    RatioMedian< 4
##    DistanceHome>=100
##    UsedChip>=0.5
##
##  Rule number: 8 [Fraud=0 cover=670892 (46%) prob=0.00]
##    RatioMedian< 4
##    DistanceHome< 100
##    DistanceLast< 50
```

Relative variable importance can be obtained from R as shown in Figure 8.4:

```
#variables importance in DT
importance <- B_DT_fraud$variable.importance
barplot(importance, main = "Decision Tree Classification-
Variable Importance", col = "lightblue", las = 2)
```

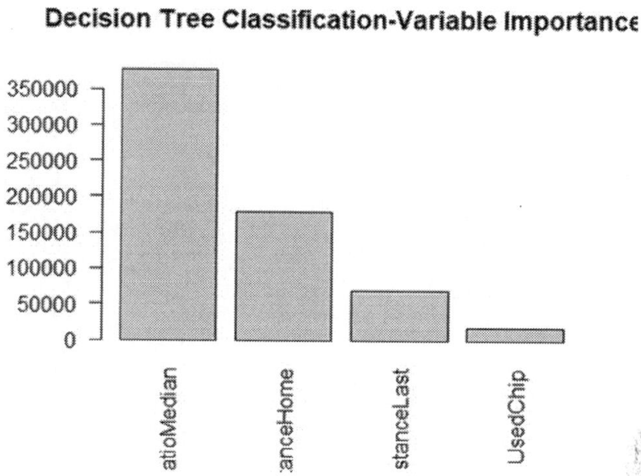

FIGURE 8.4
R Decision Tree Model Variable Importance.

Model predictions and performance statistics are obtained as follows:

```
pred_DT1 = predict(B_DT_fraud, valid.df, decision.values=
TRUE, type="class")
confusionMatrix(data=pred_DT1, reference = valid.df$Fraud,
mode = "prec_recall" , positive = "1")

## Confusion Matrix and Statistics
##
##           Reference
## Prediction      0      1
##          0 171656    432
##          1  11509 181356
##
##                Accuracy : 0.9673
##                  95% CI : (0.9667, 0.9679)
##     No Information Rate : 0.5019
##     P-Value [Acc > NIR] : < 0.0000000000000022
##
##                   Kappa : 0.9346
##
##  Mcnemar's Test P-Value : < 0.0000000000000022
##
##               Precision : 0.9403
##                  Recall : 0.9976
```

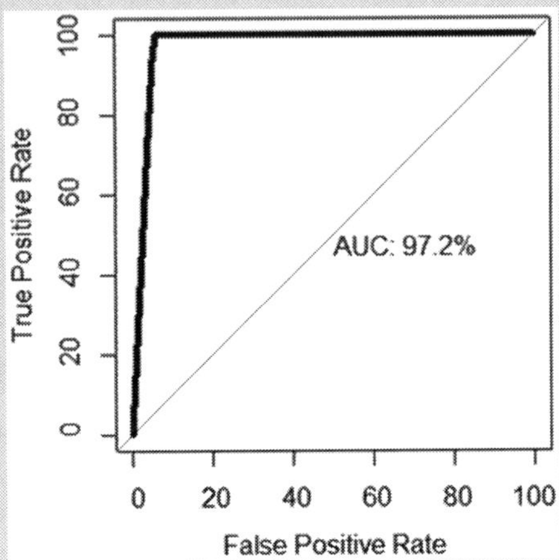

FIGURE 8.5
Decision Tree Area under the Curve.

```
##                      F1 : 0.9681
##              Prevalence : 0.4981
##          Detection Rate : 0.4969
##    Detection Prevalence : 0.5285
##       Balanced Accuracy : 0.9674
##
##        'Positive' Class : 1
##
```

Figure 8.5 displays area under the curve:

```
# ROC and AUC for DT1 model
test_prob2 = (predict(B_DT_fraud,valid.df[,-8], decision.
values= TRUE, type="prob"))[,2]
###Fix the Plot area to Square####
par(pty="s")
test_roc = roc(valid.df$Fraud ~ test_prob2, plot = TRUE,
print.auc = TRUE, legacy.axes=TRUE, percent=TRUE, xlab=
"False Positive Rate", ylab="True Positive Rate", lwd=4)

## Setting levels: control = 0, case = 1
## Setting direction: controls < cases
```

8.2.4 Both Over- and Under-Sampling Simultaneously

This process in ROSE for both under- and over-sampling was obtained by setting method="both" in uvon_sampler.

```
#2 Both Over and Undersampling while N=1000000 for data balancing
dataCredit<-as.data.frame(read.csv("CreditCardFraud.csv"))
dataCredit$Fraud= as.factor(dataCredit$Fraud)
dim(dataCredit)

## [1] 1000000       8

set.seed(123)
balanced_data=ovun.sample(Fraud~.,   data   =   dataCredit,
method = "both", N=1000000)$data
table(balanced_data$Fraud)

##
##      0      1
## 500816 499184

str(balanced_data)

## 'data.frame' :   1000000 obs. of  8 variables:
## $ DistanceHome: num  16.915 33.336 19.248 0.656 16.26 ...
## $ DistanceLast: num 0.1204 0.0157 0.5945 3.3725 0.7464 ...
## $ RatioMedian: num  0.567 1.992 0.696 1.135 2.246 ...
## $ Repeat      : int  1 1 1 0 1 1 1 0 1 1 ...
## $ UsedChip    : int  1 0 0 0 0 0 1 0 0 0 ...
## $ UsedPin     : int  0 0 0 0 0 1 0 0 0 0 ...
## $ Online      : int  1 1 0 1 1 1 0 1 0 1 ...
## $ Fraud       : Factor w/ 2 levels "0","1": 1 1 1 1 1
                  1 1 1 1 1 ...
```

This resulted in a dataset with 500,816 OK and 499,184 fraudulent cases, quite evenly balanced. The logistic regression model for this dataset using 80 percent for training and holding out 20 percent for testing is:

```
train.index      <-      sample(c(1:dim(balanced_data)[1]),
dim(balanced_data)[1]*0.8)
train.df <- balanced_data[train.index, ]
valid.df <- balanced_data[-train.index, ]
```

8.2.5 Logistic Regression with Balanced Data

```
# run logistic regression
# use glm() (general linear model) with family = "binomial" to fit a logistic
# regression.
B_fraud_model <- glm(Fraud ~ ., data = train.df, family = "binomial")
## Warning: glm.fit: fitted probabilities numerically 0 or 1 occurred

options(scipen=999)
summary(B_fraud_model)

##
## Call:
## glm(formula = Fraud ~ ., family = "binomial", data = train.df)
##
## Coefficients:
##                  Estimate Std. Error z value               Pr(>|z|)
## (Intercept)    -7.6859169  0.0251953 -305.05 <0.0000000000000002 ***
## DistanceHome    0.0290046  0.0001041  278.69 <0.0000000000000002 ***
## DistanceLast    0.0506140  0.0002674  189.27 <0.0000000000000002 ***
## RatioMedian     1.2214931  0.0031498  387.80 <0.0000000000000002 ***
## Repeat         -1.4196896  0.0147029  -96.56 <0.0000000000000002 ***
## UsedChip       -1.2125494  0.0106147 -114.23 <0.0000000000000002 ***
## UsedPin        -9.9077973  0.0740469 -133.80 <0.0000000000000002 ***
## Online          5.0128836  0.0200892  249.53 <0.0000000000000002 ***
## ---
## Signif. codes:  0 '***' 0.001 '**' 0.01 '*' 0.05 '.' 0.1 ' ' 1
##
## (Dispersion parameter for binomial family taken to be 1)
##
##     Null deviance: 1109034  on 799999  degrees of freedom
## Residual deviance:  331261  on 799992  degrees of freedom
## AIC: 331277
##
## Number of Fisher Scoring iterations: 10

# use predict() with type = "response" to compute predicted probabilities.
fraud.pred <- predict(B_fraud_model, valid.df[, -8], type = "response")

# first 10 actual and predicted records
data.frame(actual = valid.df$Fraud[1:10], predicted = ifelse(fraud.
pred[1:10]>0.5, 1,0))

##    actual predicted
## 5       0         0
## 15      0         0
## 26      0         0
## 38      0         0
## 41      0         0
## 52      0         0
## 56      0         0
```

```
## 71      0        0
## 83      0        0
## 90      0        0

####ODDS RATIOS###
round(exp(coef(B_fraud_model)),3)

## (Intercept) DistanceHome DistanceLast  RatioMedian   Repeat   UsedChip
##       0.000        1.029        1.052        3.392    0.242      0.297
##     UsedPin       Online
##       0.000      150.338

## odds ratios and 95% CI
#exp(cbind(OR = coef(fraud_model), confint(fraud_model)))
###INTERPRETATION###
##1  unit increase in predictor1, the odds of having
```

The fit of this model is obtained:

```
# Form a table from the output-- CONFUSION MATRIX ----

#install caret and use confusionMatrix()

### PLOTING AUCs#####
install.packages(c("rattle",    "rpart",    "rpart.plot",
"partykit", "caret", "pROC"), repos = "http://cran.us.r-
project.org")

## Warning: packages 'rattle', 'rpart', 'rpart.plot',
'partykit', 'caret', 'pROC'
## are in use and will not be installed

# call packages and load

library(caret)
fraud.pred= factor(ifelse (fraud.pred >0.5, 1, 0))
confusionMatrix(data=fraud.pred,   reference   =   valid.
df$Fraud, mode = "everything", positive = "1")

## Confusion Matrix and Statistics
##
##           Reference
## Prediction     0     1
##          0 93684  5218
```

```
##          1  6629 94469
##
##                Accuracy : 0.9408
##                  95% CI : (0.9397, 0.9418)
##     No Information Rate : 0.5016
##     P-Value [Acc > NIR] : < 0.00000000000000022
##
##                   Kappa : 0.8815
##
##  Mcnemar's Test P-Value : < 0.00000000000000022
##
##             Sensitivity : 0.9477
##             Specificity : 0.9339
##          Pos Pred Value : 0.9344
##          Neg Pred Value : 0.9472
##               Precision : 0.9344
##                  Recall : 0.9477
##                      F1 : 0.9410
##              Prevalence : 0.4984
##          Detection Rate : 0.4723
##    Detection Prevalence : 0.5055
##       Balanced Accuracy : 0.9408
##
##        'Positive' Class : 1
##
```

Area under the curve for this model is shown in Figure 8.6:

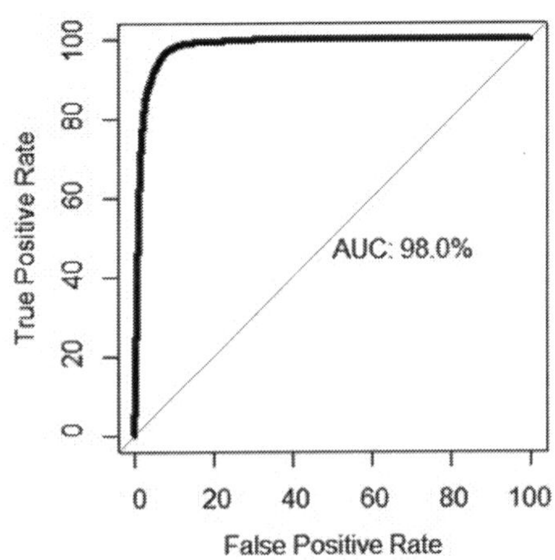

FIGURE 8.6
Area under the Curve for Logistic Regression with Balanced Data.

```
library(rpart)
library(partykit)
library(caret)
library(pROC)
library(rpart.plot)
library(rattle)

fraud.pred  <-  predict(B_fraud_model,  valid.df[,  -8],
type = "response")
#test_roc = roc(valid.df$Fraud ~ fraud.pred, plot = TRUE,
print.auc = TRUE)
###Fix the Plot area to Square####
par(pty="s")
test_roc = roc(valid.df$Fraud ~ fraud.pred, plot = TRUE,
print.auc = TRUE, legacy.axes=TRUE, percent=TRUE, xlab=
"False Positive Rate", ylab="True Positive Rate", lwd=4)

## Setting levels: control = 0, case = 1
## Setting direction: controls < cases
```

8.2.6 Decision Tree with Balanced Data

The decision tree model is applied to the training data from over-/under-sampling:

```
B_DT_fraud = rpart(Fraud ~., method="class", data=train.
df)
```

This yielded the decision tree shown in Figure 8.7:

```
#plot(as.party(DT_fraud))
prp(B_DT_fraud, type = 4, extra = 6, box.palette = "auto")
```

Error metrics were obtained, with area under the curve displayed in Figure 8.8:

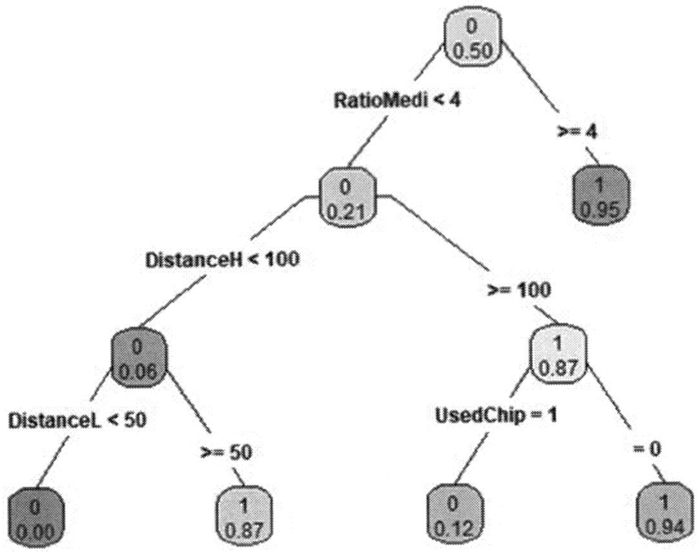

FIGURE 8.7
Decision Tree with Balanced Data.

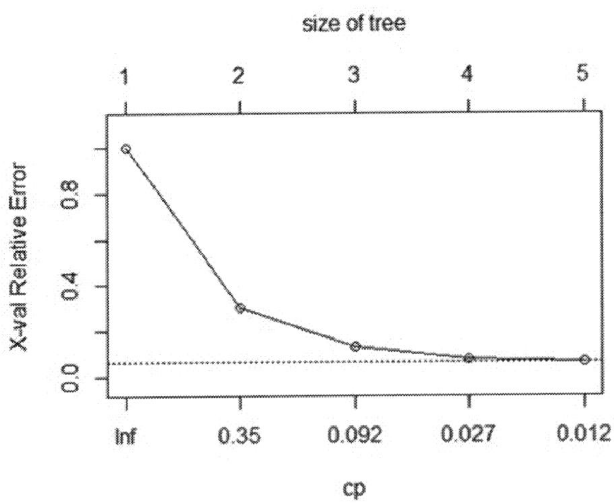

FIGURE 8.8
Decision Tree Area under the Curve for Balanced Data.

```
#fancyRpartPlot(DT_fraud)
# print the complexity parameters and plot the error vs
complexity
printcp(B_DT_fraud)

##
## Classification tree:
## rpart(formula = Fraud ~ ., data = train.df,
method = "class")
##
## Variables actually used in tree construction:
## [1] DistanceHome DistanceLast RatioMedian  UsedChip
##
## Root node error: 399497/800000 = 0.49937
##
## n= 800000
##
##           CP nsplit rel error   xerror       xstd
## 1 0.701427      0  1.000000 1.000000 0.00111944
## 2 0.169708      1  0.298573 0.298575 0.00079746
## 3 0.049387      2  0.128865 0.128867 0.00054938
## 4 0.014451      3  0.079477 0.079482 0.00043710
## 5 0.010000      4  0.065027 0.065117 0.00039711

plotcp(B_DT_fraud)
```

The decision tree rules are printed off. There are five rules, as there were with the over-sampled data:

```
#Extract the decision tree rules as text
rpart.rules(B_DT_fraud)

##  Fraud
##  0.00 when RatioMedian <  4 & DistanceHome <  100 & DistanceLast <  50
##   0.12 when RatioMedian < 4 & DistanceHome >= 100
& UsedChip is 1
##  0.87 when RatioMedian <  4 & DistanceHome <  100 & DistanceLast >= 50
##   0.94 when RatioMedian < 4 & DistanceHome >= 100
& UsedChip is 0
##  0.95 when RatioMedian >= 4

asRules(B_DT_fraud)
```

```
##
## Rule number: 3 [Fraud=1 cover=313082 (39%) prob=0.95]
##   RatioMedian>=4
##
## Rule number: 11 [Fraud=1 cover=84053 (11%) prob=0.94]
##   RatioMedian< 4
##   DistanceHome>=100
##   UsedChip< 0.5
##
## Rule number: 9 [Fraud=1 cover=26494 (3%) prob=0.87]
##   RatioMedian< 4
##   DistanceHome< 100
##   DistanceLast>=50
##
## Rule number: 10 [Fraud=0 cover=7619 (1%) prob=0.12]
##   RatioMedian< 4
##   DistanceHome>=100
##   UsedChip>=0.5
##
## Rule number: 8 [Fraud=0 cover=368752 (46%) prob=0.00]
##   RatioMedian< 4
##   DistanceHome< 100
##   DistanceLast< 50
```

Figure 8.9 shows relative variable importance:

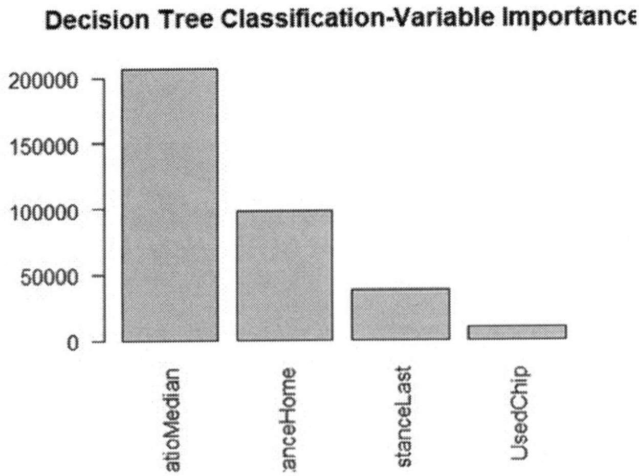

FIGURE 8.9
Relative Variable Importance for Decision Tree Area for Balanced Data.

```
#variables importance in DT
importance <- B_DT_fraud$variable.importance
barplot(importance, main = "Decision Tree Classification-
Variable Importance", col = "lightblue", las = 2)
```

We then obtain error metrics:

```
pred_DT1 = predict(B_DT_fraud, valid.df, decision.values=
TRUE, type="class")

confusionMatrix(data=pred_DT1, reference = valid.df$Fraud,
mode = "prec_recall", positive = "1")
## Confusion Matrix and Statistics
##
##           Reference
## Prediction     0     1
##          0 94094   266
##          1  6219 99421
##
##                Accuracy : 0.9676
##                  95% CI : (0.9668, 0.9683)
##     No Information Rate : 0.5016
##     P-Value [Acc > NIR] : < 0.00000000000000022
##
##                   Kappa : 0.9352
##
##  Mcnemar's Test P-Value : < 0.00000000000000022
##
##               Precision : 0.9411
##                  Recall : 0.9973
##                      F1 : 0.9684
##              Prevalence : 0.4984
##          Detection Rate : 0.4971
##    Detection Prevalence : 0.5282
##       Balanced Accuracy : 0.9677
##
##        'Positive' Class : 1
##
```

Figure 8.10 gives the area under the curve for the decision tree using under-/over-sampled data.

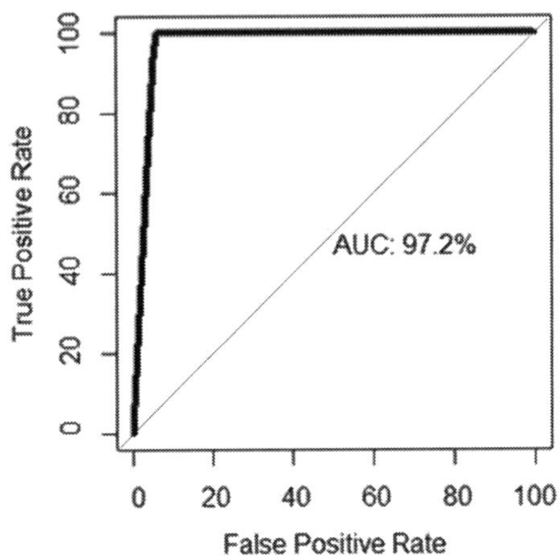

FIGURE 8.10
Area under the Curve for Decision Tree for Balanced Data.

```
# ROC and AUC for DT1 model
test_prob2 = predict(B_DT_fraud,valid.df[,-8], type=
"prob")[,2]
###Fix the Plot area to Square####
par(pty="s")
test_roc = roc(valid.df$Fraud ~ test_prob2, plot = TRUE,
print.auc = TRUE, legacy.axes=TRUE, percent=TRUE, xlab=
"False Positive Rate", ylab="True Positive Rate", lwd=4)

## Setting levels: control = 0, case = 1
## Setting direction: controls < cases
```

8.2.7 Comparisons

Precision measures how often the model correctly predicts a class, in this case fraud, so we use correct fraud cases out of all true fraud cases (true fraud divided by (true fraud plus false OK)). Recall measures how often the model finds target cases, or in our case, true fraud divided by (true fraud + false fraud)). Accuracy is the ratio of correct predictions over total predictions. Table 8.2 gives model fit measures.

The logistic regressions were slightly different although very close. The training sets were different, but the decision trees obtained were identical, testing differently as the testing sets were different. With respect to overall accuracy, the decision trees were slightly better than the logistic regression

TABLE 8.2

Classification Model Fits for Credit Card Fraud Data

	Log Regression #1	Decision Tree #1	Log Regression #2	Decision Tree #2
OK OK	171028	171656	93684	94094
ModelOK Fraud	9210	11509	5218	266
ModelFraud OK	12137	432	6629	6219
Fraud Fraud	172578	181356	94469	99421
Precision	0.9343	0.9976	0.9344	0.9411
Recall	0.9493	0.9403	0.9477	0.9973
Accuracy	0.9415	0.9673	0.9408	0.9676

models. The logistic regression models were very similar and thus recall and accuracy measures are close.

8.3 Conclusions

Classification modeling is appropriate for data with binary outcomes (as in fraud detection). These include decision trees (and related ensemble models to include random forests and extreme boosting), logistic regression, support vector machines and neural networks. Kasztelnik and Jermakowicz (2024) discussed the use of artificial intelligence in this application.

Balancing datasets is important in fraud data because it is so imbalanced by its nature. We demonstrated both over-sampling and under-/over-sampling, obtaining very similar results.

References

Alwan RH, Hamad MM, Dawood OA (2022) A comprehensive survey of fraud detection methods in credit card based on data mining techniques, *AIP Conference Proceedings* 2400, 020006.

Fevzi Esen M, Bilgic E, Basdas U (2019) How to detect illegal corporate insider trading? A data mining approach for detecting suspicious insider transactions, *Intelligent Systems for Accounting and Financial Management* 26: 60–70.

Kasztelnik K, Jermakowicz EK (2024) Financial statement fraud detection in the digital age: Artificial intelligence versus traditional methodologies, *The CPA Journal* March/April 2024: 32–39.

Siering M, Muntermann J, Grĉar M (2021) Design principles for robust fraud detection: The case of stock market manipulation, *Journal of the Association for Information Systems* 22(1): 156–178.
Vanhoeyveld J, Martens D, Peeters B (2020) Customs fraud detection: Assessing the value of behavioural and high-cardinality data under the imbalanced learning issue, *Pattern Analysis and Applications* 23: 1457–1477. https://doi.org/10.1007/s10044-019-00852-w.
Velasco RB, Carpaneses I, Interian R, Paulo Neto OCG, Ribeiro CC (2021) A decision support system for fraud detection in public procurement, *International Transactions in Operational Research* 28: 27–47. doi:10.1111/itor.12811.
Yang Y, Xu Y, Sun Y, Dong Y, Wu F, Zhuang Y (2021) Mining fraudsters and fraudulent strategies in large-scale mobile social networks, *IEEE Transactions on Knowledge and Data Engineering* 33(1): 169–179.

9

Mixed Data

This chapter deals with data abnormalities that are commonly incurred. Especially with data involving fraud, the hope is that the fraud cases are relatively rare, which leads to imbalanced datasets (very few fraud cases compared to those that are not). This often leads to models simply classifying all cases as not fraud, obtaining very high accuracy rates. There are well-developed means to deal with this problem. Preprocessing techniques include under-sampling (which makes computation easier, but reduces data content), over-sampling (which increases computational burden). One over-sampling method is SMOTE (synthetic minority over-sampling technique). SMOTE randomly draws observations from the smaller set of outcomes, which can provide exactly the degree of balance desired, but risks some unintentional bias. In Chapter 8 we used the R ROSE over-sampling method, which is very similar in operation to SMOTE.

9.1 Big Data Mining

The age of big data arose because of the ability to tap massive streams of data. Within an enterprise, most firms now have enterprise resource planning (ERP) systems, which usually include customer relationship management (CRM) tools. Both sources provide streams of internal data available in real time. Data comes in a variety of structures. Transactional data and data from electronic data interchange (EDI) have some structure but require some processing to put into form needed by data mining software. There are also a variety of sources of unstructured data, to include documents, videos, machine-sensed inputs from production operations, and data from social media.

Once data is obtained and processed, there are many tools available for data visualization and data analytics. Visualization is supported by software tools such as Tableau and other forms of real-time dashboards. Data analytic tools range from statistical analysis, optimization, association rules, clustering, and classification.

DOI: 10.1201/9781003562399-9

9.2 Categorical Data

Some data mining algorithms assume data normality, which often is not the case. Regression is affected – if the outcome variable is binary, logistic regression is appropriate. But often the independent variables are binary or categorical. The Employee Attrition Classification dataset is a case in point.

Taha and Hadi (2019) addressed categorical data in data mining.

9.2.1 Employee Attrition Classification Dataset

The Synthetic Employee Attrition Dataset is a simulated dataset designed for the analysis and prediction of employee attrition. It contains detailed information about various aspects of an employee's profile, including demographics, job-related features, and personal circumstances.

The dataset comprises 74,498 samples, split into training (n=59598: 31200 stay, 28338 left) and testing (n=14900: 7868 stay, 7032 left) sets to facilitate model development and evaluation. Each record includes a unique Employee ID and features that influence employee attrition. The goal is to understand the factors contributing to attrition and develop predictive models to identify at-risk employees.

This dataset is ideal for HR analytics, machine learning model development, and demonstrating advanced data analysis techniques. It provides a comprehensive and realistic view of the factors affecting employee retention, making it a valuable resource for researchers and practitioners in the field of human resources and organizational development.

9.2.2 Features

Age: The age of the employee, ranging from 18 to 60 years.

Gender: F0M1 Female 0; Male 1.

YrCo: The number of years the employee has been working at the company.

MoInc: The monthly salary of the employee, in dollars.

WLBal: Employee's perceived balance between work and personal life (Poor 0, Below Average 1, Good 2, Excellent 3).

Sat: The employee's satisfaction with their job (Very Low 0, Low 1, Medium 2, High 3).

Perf: The employee's performance rating (Low, Below Average, Average, High).

Number of Promotions: The total number of promotions the employee has received.

OT: No 0; Yes 1.

DistHome: The distance between the employee's home and workplace, in miles.

Ed: The highest education level attained by the employee (High School 0, Associate Degree 1, Bachelor's Degree 2, Master's Degree 3, PhD 4).

Marital: Employee marital status (Divorced 2, Married 1, Single 0).

Depend: Dependents.

JobLevel: The job level of the employee (Entry 0, Mid 1, Senior 2).

CoSize: The size of the company the employee works for (Small 0, Medium 1, Large 2).

TenMos: The total number of years the employee has been working in the industry.

Remote: Whether the employee works remotely: (Yes 1 or No 0).

LeadOp: Whether the employee has leadership opportunities: (Yes 1 or No 0).

InnOp: Whether the employee has opportunities for innovation: (Yes 1 or No 0).

CoRep: Employee perception of company reputation: (Very Poor 0, Poor 1,Good 2, Excellent 3).

Reco: The level of recognition the employee receives (Very Low 0, Low 1, Medium 2, High 3).

Attrition: Whether the employee has left the company, encoded as 0 (Stayed) and 1 (Left).

Table 9.1 displays correlations with the target Attrition. Table 9.2 gives details of the outcomes in terms of Attrition for each value of each variable.

9.3 Naïve Bayes

Another classification method is Naïve Bayes modeling. It has been applied by Apple to assess external risk factors in their supply chain network (Lockamy 2017). Apple's international suppliers were assessed on six risk factors, and a Naïve Bayes model generated potential risk probabilities.

Naïve Bayes classifier provides a simple means of classifying categorical data. It seeks to model each class of the training data by probability density functions. Items are assigned to the most probable class. The method assumes a number of explanatory variables and assumes them to be independent. It

TABLE 9.1

Correlation with Outcome

	Attrition
Age	–0.049
FOM1	–0.102
YrCo	–0.065
MoInc	–0.011
WLBal	–0.184
Sat	–0.064
Perf	–0.052
Promos	–0.081
OT	0.056
DistHome	0.096
Ed	–0.045
Marital	–0.230
Depend	–0.078
JobLevel	–0.314
CoSize	–0.026
TenMos	–0.030
Remote	–0.220
LeadOp	-0.010
InnOp	–0.022
CoRep	–0.085
Recog	–0.001

TABLE 9.2

Training Dataset Counts

	Stay	Leave	Prob(Stay)	Prob(Leave)
Age 0	5890	6734	0.466534	0.533466
Age 1	5971	5302	0.529626	0.470374
Age 2	7115	5851	0.548797	0.451203
Age 3	6102	5303	0.535051	0.464949
Age 4	6182	5148	0.546107	0.453893
FOM10	12580	14279	0.468312	0.531688
FOM11	18680	14059	0.570811	0.429189
YrCo0	5615	6383	0.468279	0.531721
YrCo1	5745	6118	0.484141	0.515859
YrCo2	6601	5452	0.547853	0.452147
YrCo3	7132	5521	0.56369	0.43631
YrCo4	6167	4864	0.559069	0.440931
MoInc0	6093	5835	0.511016	0.488984
MoInc1	6253	5661	0.524712	0.475288
MoInc2	6356	5569	0.532801	0.467199
MoInc3	6232	5686	0.522955	0.477045
MoInc4	6326	5587	0.531553	0.468447
WLBal0	3313	4992	0.398916	0.601084
WLBal1	7676	10370	0.425357	0.574643
WLBal2	13391	9137	0.594416	0.405584

TABLE 9.2 (Continued)

Training Dataset Counts

	Stay	Leave	Prob(Stay)	Prob(Leave)
WLBal3	6880	3839	0.641851	0.358149
Sat0	5706	6405	0.471142	0.528858
Sat1	2803	3088	0.475811	0.524189
Sat2	6461	5356	0.546755	0.453245
Sat3	16290	13489	0.54703	0.45297
Perf0	1253	1697	0.424746	0.575254
Perf1	4313	4637	0.481899	0.518101
Perf2	19283	16527	0.538481	0.461519
Perf3	6411	5477	0.539283	0.460717
Promo0	15058	14658	0.50673	0.49327
Promo1	7629	7336	0.50979	0.49021
Promo2	5543	5341	0.50928	0.49072
Promo3	2422	823	0.746379	0.253621
Promo4	608	180	0.771574	0.228426
OT0	21835	18313	0.543863	0.456137
OT1	9425	10025	0.484576	0.515424
Dist0	6965	4989	0.58265	0.41735
Dist1	7012	5070	0.580367	0.419633
Dist2	6289	5894	0.516211	0.483789
Dist3	5694	6338	0.473238	0.526762
Dist4	5300	6047	0.467084	0.532916
Ed0	6080	5668	0.517535	0.482465
Ed1	7685	7230	0.515253	0.484747
Ed2	9035	8791	0.506844	0.493156
Ed3	6139	5881	0.510732	0.489268
Ed4	2321	768	0.751376	0.248624
Marit0	6912	13923	0.331749	0.668251
Marit1	19097	10811	0.638525	0.361475
Marit2	5251	3604	0.592998	0.407002
Depend0	8968	8880	0.502465	0.497535
Depend1	7832	7678	0.504965	0.495035
Depend2	4425	4437	0.499323	0.500677
Depend3	4092	4157	0.49606	0.50394
Depend4	3946	2120	0.650511	0.349489
Depend5	1578	823	0.657226	0.342774
Depend6	419	243	0.632931	0.367069
Job0	8785	15082	0.368081	0.631919
Job1	12930	10823	0.544352	0.455648
Job2	9545	2433	0.796878	0.203122
CoSize0	8990	8945	0.501255	0.498745
CoSize1	15890	13855	0.534207	0.465793
CoSize2	6380	5538	0.535325	0.464675
Ten0	6120	5967	0.506329	0.493671
Ten1	6360	5910	0.518337	0.481663
Ten2	6284	5697	0.524497	0.475503
Ten3	6354	5724	0.52608	0.47392
Ten4	6142	5040	0.549276	0.450724
Remote0	22725	25514	0.471092	0.528908

(Continued)

TABLE 9.2 (Continued)

Training Dataset Counts

	Stay	Leave	Prob(Stay)	Prob(Leave)
Remote1	8535	2824	0.751387	0.248613
Lead0	29664	27016	0.523359	0.476641
Lead1	1596	1322	0.54695	0.45305
Innov0	25934	23961	0.519772	0.480228
Innov1	5326	4377	0.548902	0.451098
CoRep0	5679	6138	0.480579	0.519421
CoRep1	5235	6799	0.435017	0.564983
CoRep2	16992	12774	0.570853	0.429147
CoRep3	3354	2627	0.560776	0.439224
Recog0	1596	1367	0.538643	0.461357
Recog1	12395	11363	0.521719	0.478281
Recog2	9429	8604	0.522875	0.477125
Recog3	7840	7004	0.52816	0.47184

is considered to have the relative advantage that it does not require a large training dataset for effective classification. Naïve Bayes is a conditional probability model. For a vector of n factors (independent variables), for K possible outcomes or classes, you seek:

$$Prob\left\{C_k \mid x_1, x_2, \ldots x_n\right\}$$

It would be accurate to enumerate all combinations of classes, but this can become unwieldy if n is large, and in the simple example to follow, the sample size might be quite low. Naïve Bayes offers a simple alternative. Bayes theorem can be applied to estimate this probability:

$$Prob\left\{C_k \mid x_1, x_2, \ldots x_n\right\} = \frac{Prob\left\{C_k\right\} Prob\left\{x_1, x_2, \ldots x_n\right\} \mid C_k\}}{Prob\left\{x_1, x_2, \ldots x_n\right\}}$$

Restating this,

$$Posterior = \frac{prior \times likelihood}{evidence}$$

The denominator doesn't depend on C_k, and the vector of independent variable values $x_1, x_2, \ldots x_n$ are given, to the denominator is a constant. A formal statement of the method is:

Probability{Membership in Class *k*} is given by the formula:

$$Prob\{C_k\}\prod_{i=1}^{n}Prob\{x_i|C_k\}$$

9.3.1 Employee Attrition Dataset with Categorical Data

Of the training set, Table 9.3 gives an enumeration of the two ages, two incomes, two risk categories, and three credit categories, as well as counts of OK and Problem.

To simplify the Naïve Bayes model we focus on those variables that have differences in probability of staying of at least 0.05. This also allows combination for quite a few variables. Those left are given in Table 9.3.

TABLE 9.3

Selected Variable Data

Variable	Range	Count Stay	Count Leave	Prob{Stay}	Prob{Leave}
Age	<27	5890	6734	0.467	0.533
	27+	25370	21604	0.540	0.460
F0M1	Female	12580	14279	0.468	0.532
	Male	18680	14059	0.571	0.429
YrCo	<11 years	11360	12501	0.476	0.524
	11 years +	19900	15837	0.557	0.443
WLBal	0	3313	4992	0.399	0.601
	1	7676	10370	0.425	0.575
	2	13391	9137	0.594	0.406
	3	6880	3839	0.642	0.358
Sat	0 or 1	8509	9493	0.473	0.527
	2 or 3	22751	18845	0.545	0.453
Perf	0	1253	1697	0.425	0.575
	1	4313	4637	0.482	0.518
	2 or 3	25694	22004	0.539	0.461
Promo	0, 1 or 2	28230	27335	0.508	0.492
	3 or 4	3030	1003	0.751	0.249
OT	0	21835	18313	0.544	0.456
	1	9425	10025	0.485	0.515
DistHome	< 41 miles	13977	10059	0.582	0.418
	41-60 miles	6289	5894	0.516	0.484
	Over 60 miles	10994	12385	0.470	0.530
Ed	< PhD	28939	27570	0.512	0.488
	PhD	2321	768	0.751	0.249
Marital	Single	6912	13923	0.332	0.668
	other	24348	14415	0.628	0.372
Dependents	0, 1 or 2	21225	20995	0.503	0.497
	Over 2	10035	7343	0.577	0.423

(Continued)

TABLE 9.3 (Continued)

Selected Variable Data

Variable	Range	Count Stay	Count Leave	Prob{Stay}	Prob{Leave}
Job	0	8785	15082	0.368	0.632
	1	12930	10823	0.544	0.456
	2	9545	2433	0.797	0.203
Remote	no	22725	25514	0.471	0.523
	yes	8535	2824	0.751	0.249
CompRep	0 or 1	10914	12937	0.458	0.542
	2 or 3	20346	15401	0.569	0.431

Bayesian probability would be:

$$Prob\left\{x_i, 1=1,n\right\} = \frac{\prod_{i-1}^{n} x_i}{\prod_{i=1}^{n} x_i + \prod_{i=1}^{n} \left(1-x_i\right)}$$

9.3.2 Naïve Bayes with R

R package "e1071" can be used for building Naïve Bayes classifiers. To run a Naïve Bayes model with the employee attrition data we do the following:

```
#Model1 - Naive Bayes
TrainData = as.data.frame(read.csv("EmpAttTrainCat.csv"))
TestData = as.data.frame(read.csv("EmpAttTestCat.csv"))
#select variables to reduce the data
model_var <- c(6,9,12,13,15,18, 23)

TrainData=as.data.frame(TrainData[1:nrow(TrainData),
model_var])
TestData=as.data.frame(TestData[1:nrow(TestData), model_var])
#variable conversion
TrainData$Left=as.factor(TrainData$Left)

TestData$Left=as.factor(TestData$Left)

# call packages and load
install.packages('e1071', repos = "http://cran.us.r-proj
ect.org")

library(e1071)
install.packages(c("rpart", "rpart.plot", "partykit",
"caret", "pROC"), repos = "http://cran.us.r-project.org")
```

The following accesses the training dataset and calculates imputed probabilities:

```
library(caret)
## Loading required package: ggplot2
## Loading required package: lattice
library(pROC)
library(partykit)
library(rpart.plot)
## Loading required package: rpart
NB_model <- naiveBayes(TrainData$Left ~ ., data = TrainData)
NB_model
```

The Naïve Bayes model is accessed as follows:

```
## Naive Bayes Classifier for Discrete Predictors
##
## Call:
## naiveBayes.default(x = X, y = Y, laplace = laplace)
##
## A-priori probabilities:
## Y
##         0         1
## 0.5245142 0.4754858
##
## Conditional probabilities:
##    WLBal
## Y       [,1]      [,2]
##   0 1.762572 0.9128121
##   1 1.417214 0.9308271
##
##    Promos
## Y        [,1]      [,2]
##   0 0.9089251 1.0657052
##   1 0.7483591 0.9032013
##
##    Ed
## Y       [,1]     [,2]
##   0 1.710045 1.197156
##   1 1.606571 1.102898
```

```
##
##    Marital
## Y          [,1]      [,2]
##   0 0.946865 0.6215147
##   1 0.635860 0.6970777
##
##    JobLevel
## Y          [,1]      [,2]
##   0 1.0243122 0.7653757
##   1 0.5536382 0.6471867
##
##    Remote
## Y            [,1]      [,2]
##   0 0.27303263 0.4455246
##   1 0.09965417 0.2995436
```

Prediction is accomplished here for six initial training observations:

```
pred.prob <- predict(NB_model, newdata = TestData,
type = "raw")
pred.class <- predict(NB_model, newdata = TestData)
predictions <- data.frame(actual = TestData$Left,
predicted = pred.class, pred.prob)
head(predictions)
##   actual predicted        X0         X1
## 1      0         0 0.6076674 0.39233256
## 2      1         1 0.1402005 0.85979947
## 3      0         0 0.9583660 0.04163402
## 4      1         1 0.2205760 0.77942398
## 5      0         0 0.9656312 0.03436878
## 6      0         0 0.8736327 0.12636726
```

In this case the model was accurate in all cases. Strength of confidence can be assessed by the probabilities given for each output. The fit of this model is obtained as follows:

```
library(caret)
#Confusion Matrix
pred.class <- predict(NB_model, newdata = TestData)
```

```
confusionMatrix(pred.class, TestData$Left, positive = "1",
mode = "everything")

## Confusion Matrix and Statistics
##
##           Reference
## Prediction    0    1
##          0 5927 2498
##          1 1941 4534
##
##                Accuracy : 0.7021
##                  95% CI : (0.6947, 0.7094)
##     No Information Rate : 0.5281
##     P-Value [Acc > NIR] : < 2.2e-16
##
##                   Kappa : 0.3998
##
##  Mcnemar's Test P-Value : < 2.2e-16
##
##             Sensitivity : 0.6448
##             Specificity : 0.7533
##          Pos Pred Value : 0.7002
##          Neg Pred Value : 0.7035
##               Precision : 0.7002
##                  Recall : 0.6448
##                      F1 : 0.6714
##              Prevalence : 0.4719
##          Detection Rate : 0.3043
##    Detection Prevalence : 0.4346
##       Balanced Accuracy : 0.6990
##
##        'Positive' Class : 1
##
```

This model was 70 percent accurate, with sensitivity of 0.64 and specificity of 0.75. The area under the curve can be obtained as follows:

```
par(pty="s")
test_prob = (predict(NB_model, TestData, type = "raw"))
[,2]
test_roc = roc(TestData$Left ~ test_prob, plot = TRUE,
print.auc = TRUE, legacy.axes=TRUE, percent=TRUE,
```

```
xlab="False Positive Rate", ylab="True Positive
Rate", lwd=4)
## Setting levels: control = 0, case = 1
## Setting direction: controls < cases
#####AUC##############
auc(test_roc)
## Area under the curve: 76.3%
###Return back to Maximum Plot Area###
par(pty="m")
```

The plot of the area under the curve is given in Figure 9.1.

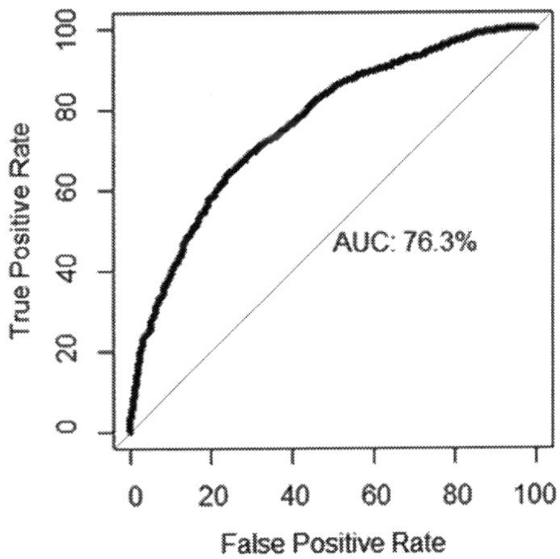

FIGURE 9.1
Area under the Curve for the R Naive Bayes Model.

9.4 Employee Attrition Dataset with Continuous Data

The file EmployeeAttritionTrainClean.csv consists of 22 variables, one of which was staying or leaving. The R decision tree model is built with:

```
dt <- rpart(TrainData$Left ~ ., data = TrainData, method=
"class")

#plot(as.party(dt))
prp(dt, type = 4, extra = 6, box.palette = "auto")
```

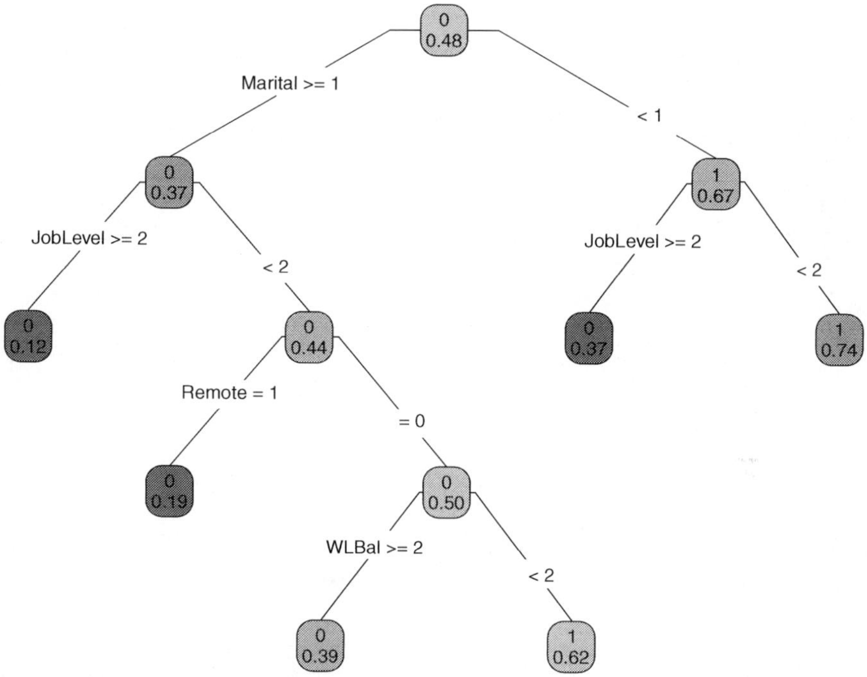

FIGURE 9.2
R Decision Tree for EmployeeAttritionTrainClean Model.

Figure 9.2 displays the decision tree model obtained using R.

This model ended up with six rules. Figure 9.3 displays the decrease in relative error versus the number of tree rules.

```
plotcp(dt)
```

Rule details can be obtained as follows:

```
rpart.rules(dt)
##  TrainData$Left
##            0.12 when Marital >= 1 & JobLevel >= 2
##            0.19 when Marital >= 1 & JobLevel <  2
& Remote is 1
##            0.37 when Marital <  1 & JobLevel >= 2
##            0.39 when Marital >= 1 & JobLevel <  2
& Remote is 0 & WLBal >= 2
```

```
##             0.62 when Marital >= 1 & JobLevel <  2 &
Remote is 0 & WLBal <  2
##             0.74 when Marital <  1 & JobLevel <  2
```

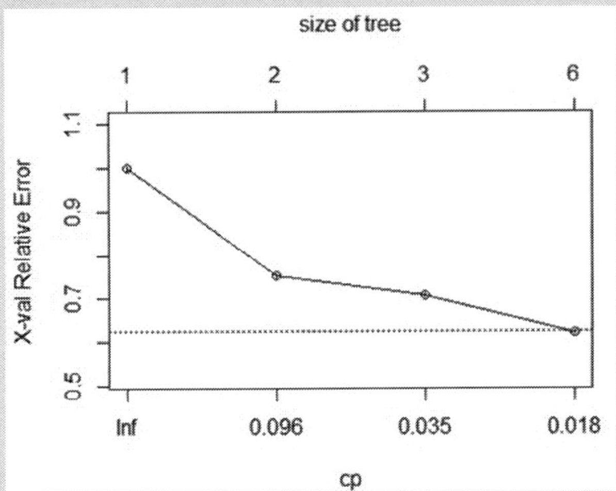

FIGURE 9.3
Decision Tree Model Relative Error versus Tree Size.

The model consists of six rules. Model fit is called in R:

```
DT_pred <- predict(dt, TestData, decision.values=TRUE,
type="class")
confusionMatrix(data=DT_pred, reference = TestData$Left,
mode = "everything", positive = "1")

## Confusion Matrix and Statistics
##
##           Reference
## Prediction    0    1
##          0 6475 3430
##          1 1393 3602
##
##                Accuracy : 0.6763
##                  95% CI : (0.6687, 0.6838)
##     No Information Rate : 0.5281
##     P-Value [Acc > NIR] : < 2.2e-16
```

```
##
##                  Kappa : 0.3404
##
## Mcnemar's Test P-Value : < 2.2e-16
##
##            Sensitivity : 0.5122
##            Specificity : 0.8230
##         Pos Pred Value : 0.7211
##         Neg Pred Value : 0.6537
##              Precision : 0.7211
##                 Recall : 0.5122
##                     F1 : 0.5990
##             Prevalence : 0.4719
##         Detection Rate : 0.2417
##   Detection Prevalence : 0.3352
##      Balanced Accuracy : 0.6676
##
##       'Positive' Class : 1
##
```

Model accuracy is 68 percent with sensitivity 0.51 and specificity 0.82. Model area under the curve is obtained and plotted in Figure 9.4:

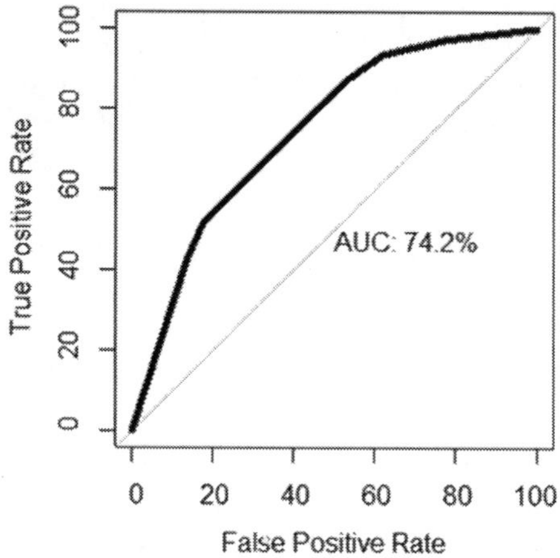

FIGURE 9.4
Area under the Curve for the Decision Tree Model.

```
par(pty="s")
test_prob    =   predict(dt,    TestData,    decision.values=
TRUE,type = "prob")[,2]
test_roc = roc(TestData$Left ~ test_prob, plot = TRUE,
print.auc = TRUE, legacy.axes=TRUE, percent=TRUE, xlab=
"False Positive Rate", ylab="True Positive Rate", lwd=4)

## Setting levels: control = 0, case = 1
## Setting direction: controls < cases

#####AUC#############
auc(test_roc)
## Area under the curve: 74.24%
```

This chapter demonstrated the use of Naïve Bayes modeling on categorical data, as well as comparing a decision tree model using continuous data. Results are not strictly comparable as they are based on different variables. The Naïve Bayes model provided a slightly better fit here. Other classification models probably would better that fit.

9.5 Conclusions

Naïve Bayes modeling provides a tool to deal with ordinal data. However, it is often less accurate than classification models assuming the data to be ratio (quantitative). While statistically a problem, when dealing with ordinal data, it is best to be selective in picking variables for modeling.

References

Lockamy, A (2017) An examination of external risk factors in Apple Inc.'s supply chain, *Supply Chain Forum* 18(3): 177–188.

Taha A, Hadi AS (2019) Anomaly detection methods for categorical data: A review, *ACM Computing Surveys* 52(2): 1–35. Article 38.

Vanhoeyveld J, Martens D, Peeters B (2020) Customs fraud detection: Assessing the value of behavioural and high-cardinality data under the imbalanced learning issue, *Pattern Analysis and Applications* 23: 1457–1477. https://doi.org/10.1007/s10044-019-00852-w

Index

Note: Page numbers in *italics* indicate figures and in **bold** indicate tables on the corresponding pages.

FSC
www.fsc.org
MIX
Papier | Fördert
gute Waldnutzung
FSC® C083411

Kolibri 360 GmbH
Ferdinand-Jühlke-Straße 7
99095 Erfurt, Deutschland
produktsicherheit@kolibri360.de